W0082200

During the last two decades, as a result of the development of information technology as the primary tool of architects and engineers, the convergence between disciplines is creating exciting new potentials for both practices, and a far more intimate relationship between them.

Design Engineering is rapidly becoming a creative tool, and an intrinsic part of a new generation of form and organisation. The resulting opportunities constitute an integral component of architectural thought.

This book spans Adams Kara Taylor's first decade of production. Its structure and organisation follows that of an unrealised exhibition project on the office's conceptual work. The following pages are thus a journey through the ideas and research that produce AKT's work, rather than a chronological or typological inventory of the work itself. This itinerary is guided by transcripts of conversations between Actar and AKT.

Design Engineering

AKT

Adams
Kara
Taylor

Table of contents

Table of icons

EXTREMES

UNLEARNING

INTERSECTION

PROCESS

In the traditional relationship between architecture and engineering, the architect has been the creator and the engineer the problem-solver. AKT operates somewhere between these two roles, adding something to the architect's vision to make it even better. Each architect requires a different process, and perhaps a different relationship with the engineer to advance the project. Although each works differently they all require AKT to operate not just as problem-solvers, but as contributors.

In this process, the engineer can often suggest ideas that push an architect's practice forward creating a feedback loop in which architectural problems about engineering solutions can then inform future practice. AKT's approach requires an empathy with both the architectural project and the architect's process. In each case it also requires an ability to see the project on the architect's terms, whilst using design engineering to play a crucial role in helping better realize these concepts.

If traditionally architecture and engineering had been based upon a reversible, linear and hierarchical correspondence between the organisation and the production process, then parametric design (a technology enabled by computing) is now increasingly relegating the final vision to be a product of a construction process capable of incorporating a far more complex set of considerations than the traditional architectural premises were able to integrate. This more concrete and non-idealistic design approach, developed by engineering technologies, is producing a radically different approach to design.

Re Design Engineering

Simon Allford

Founding partner of Allford Hall Monaghan Morris, lecturer at the Bartlett School of Architecture

ENGINEERING ACTIONS AND REACTION
Actions and Reactions define architectural propositions and therefore the lot of the architect, who conjures up designs that can be communicated as comment, text, process, diagram, drawing and even buildings. The invention of the profession has increased the frequency with which designs become proposals. Professions also diversify and spawn new ones. So, what about the structural engineer? Engineering is a reaction to architecture. Architects need good clients and engineers need good architects: historically that is their lot.

HISTORICAL REACTION AND A PARTICULAR CONTEXT
Engineers increasingly question this collaboration. The history they cite is littered with design failure. They point out, correctly, that neither design nor construction is exclusively owned by architecture or architects. London and its architects however have boasted of 'their' engineers since Brunel, here the Corbusian fascination with the magnificent engineer lives on. Engineers profit from and indulge in this adulation that also keeps them in their collaborative place. Ove Arup, initially as a contractor and Samuely, his engineer, questioned each other's roles and superiority. Later Newby, Rice, Hunt and Happold traded staff, ideas and even occasional insults. Current debates only repeat old arguments. Time, design's shrewdest critic, will tell.

So how does an engineer, specifically Adams Kara Taylor, operate in this murky world? The engineers chose, as outlined in the many conversations with Hanif Kara, to develop engineering inspired either by the pursuit of material properties or by collaboration with architects. Surprisingly, to those aware of their considerable design reputation and their penchant for exploration, they chose the latter. Or so they believe, or would have us believe. Perhaps what I think I know is less important than what they encourage and facilitate.

RE DESIGN PRACTICE
London does not survive on engineering mythology alone: there is more to design than architecture, more to architecture than structure. AKT's rapid rise has coincided with many arguments about Information Technology's potential to transform the process of architectural design and therefore architecture. They have been involved in this debate about the use of computer power to process information and it has informed who they work with and by what means.

They have used this debate and this technology well, precisely because they are aware of the fallacies. They understand that time is compressed and that a lot more information can be produced more rapidly. But they also know that IT's ever increasing capacity can propagate only the process and that generating process does not itself generate ideas. They know that it is a mistake to assume that the design process is itself changing. They are smarter than some critics who mistake mass production for mass invention. They are wary and suspect, as I do, that the current penchant for processing facilitated by the speed of the computer is in fact as likely to reinvent the architecture of the Renaissance as it is the architecture of Nanotechnology, especially as the construction systems of both remain a mystery.

Throughout this discussion AKT have remained engaged but critical. Through activities at the Architectural Association they (Hanif) have become academic speculators on design theory; whilst in commercial practice, they have built speculations on contemporary construction. More importantly, this twin track approach has saved them from the encryptions and prejudices of the all too often divided worlds of academe and practice. AKT have built bridges across the artificial divide because they know that the real world is the world of architectural ideas: both on paper and on site.

CONSTRUCTING THE IDEA
There are similarities in how both our offices practice. We share many beliefs. We believe that design is understanding the determining factors that combine to create any situation and reacting to them by rethinking their juxtaposition and composition. That in the most extreme cases design means starting afresh and reorganising, but also that this indefatigable search for the new and innovative can be, at times, foolish. To commence design by jettisoning all is a radical action but nothing more and that, too much of what passes as design analysis is just process presented as product. The pursuit of difference, as an end in itself, is to confuse design innovation with design conceit: the future of design is what it always has been as the new continues to shock in a very familiar way. I think Hanif and I agree about all the above, but I am not sure. In the end AKT are consultant engineers!

CONSTRUCTING PRACTICE MODELS
My own experience of working closely with AKT's leading designers is that there is a shared culture of defining matrices of opportunities and an almost ritual questioning of the same. It is all very simple as design can be. It is also never certain or comfortable: complexity is readily available and does not need to be sorted out or even more perversely invented. Indeed invented complexity is the most banal yet frequent outcome of the worst excesses of design facilitated by IT. In an unfettered world of

technological awe, ideas about architecture become 'cultural production', whatever that may be. The ugly juxtaposition of the words alone confirms that I don't need it.

In conversation there is, inevitably (they are too worldly and well travelled to be immune), the occasional reference to 'parametric design', 'design intelligence' and the new 'complex' calculations that can be undertaken. This, thankfully, is countered by questioning why they should be undertaken, how cost can be rethought and funded (the latter is an interesting shift in any definition of the design parameters of engineering) and the realities of the sub contractor world, even if said 'subbie' is actually from another field of manufacture.

Knowing that I am interested in setting up systems that discover rather than determine potential forms, my own experience of AKT is peculiar to how they wish to present themselves to me. This began with a lunch with Hanif when KaraTaylor opened; Albert was out solving problems and Robin Adams was yet to join them, though they would have us believe the master plan was already in place. It took a few years to get a project together but success in competitions ensured that eventually the work flowed. I soon learned that this was to be an all embracing collaboration: everything from idea to detail was to be challenged. This suited the exploratory processes I employ but was unusual for an engineer: all too often confronted by process they retreated into the art of engineering (an inbred resistance to confinement?). In our collaboration the rigorous architectural process has been confronted by the engineering equivalent. Assumptions, smart or bland, are deemed idol and subjected to critical analysis. Who were we competing against with both intellectually and personally informed strategies? These strategies are not necessarily strategies of design but strategies of design of how to win. As we began working more with clients I noticed that they too were subject to the most severe form of assessment: the more the client knew the more rigorous the examination.

Of course as the projects changed so have the challenges. CASPAR was a smart idea about a new apartment typology with a clever structural idea about bridges spanning onto bridges. Ironically with First Base what began with the invention of a prototype has ended up as a series of very different projects. Initial explorations of the benefit of a standard (ish) concrete frame has led to courtyard housing; the study of how two existing towers could be extended and re-invented using both light and heavy cladding, and how to build a thirty story tower on a sliver of land between two transport infrastructure projects (one a tunnel to Rotherhithe the other a link to Limehouse). We are still setting up the rules for the Olympic Village master plan. We have collaborated long enough to acknowledge early on that any project based on an idea of repeat systems was unlikely to repeat them: changing circumstances dictate design. Union Square, with the maverick architect client Roger Zogolovitch, is therefore a 'bespoke-exemplar': exploring a master plan for a fragment of city and then building it as a critique of current construction, regulatory and contractual frameworks all under a newly invented fee regime.

It is however at Notting Hill where the benefits of collaborating with AKT have had the most impact on the architecture. Monsoon HQ, Phase One, is the result of our shared interest in how we might make a concrete frame and a tradition of the speculative office work harder. The design was solved by inventing and communicating the benefits of a hard won and hard working 'decorative' structure. The speculative office in Phase Two achieves what we could not achieve in Phase One, an external structure: appropriately in contrast this resists the desire to be 'decorative' as it doubles up the solar shade called into being by regulations absent in Phase One. Our history is very much one of design as the redefinition of responses to ever changing contexts; design as both style and a modus operandi. And of course it moves on: at Blackfriars the two pursuits are combining to create a taller related structure, at Angel to salvage a frame.

DESIGN ENGINEERING

I know Hanif has a particular idea about the title of this publication 'Design Engineering'. Design comes before Engineering because AKT is driven firstly by design as a pure problem solving discipline and secondly by engineering as the particular professional discipline. He is also well aware of current pre-occupations with shapes; how the changing cultural context redefines trends and how the current quasi-religious fascination with design information processing is subject to the ebb and flow of fashion. He is also interested in people and business and politics. AKT is interested in design, engineering, the engineering of design and design engineering.

FINDING FORM NOT FORM FINDING

I was flattered to be asked to make these observations but then recalled a recent public debate which confirmed that the engineer's reactions are changing: Hanif, in typically truculent form remarked 'I don't need to be an engineer when discussing ideas with Simon because he's a bloody good engineer... but then again he needs to be- because I am a bloody good architect'!

Located adjacent to the Town Square, the Peckham Library and Media Centre was commissioned in 1995 by Southwark Borough Council as an arts and technology hub. In order to provide a public amenity, the main reading room is raised twelve meters into the air to create a covered public plaza on the ground level, and to give library readers an elevated view of the city. Taking on the challenge of supporting a reading room full of books above a long span, AKT's solution uses concrete-framed lower floors containing the entrance and lift to support one side of the reading room. The other side rests on seven slender, inclined columns that pierce the reading room floor and continue to the roof. The angle of incline of these columns both eliminates the need for separate cross-bracing, and provides a structural solution that animates, rather than disrupts, the ground plane below. Steel structure is used for the long-span reading room and the sloping external columns, whereas wood-frame construction is used for a series of "pods" suspended within the reading room.

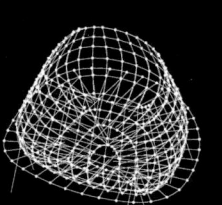

PODS
PECKHAM LIBRARY

ALSOP ARCHITECTS
PECKHAM, LONDON, 1999
CLIENT: LONDON BOROUGH OF SOUTHWARK

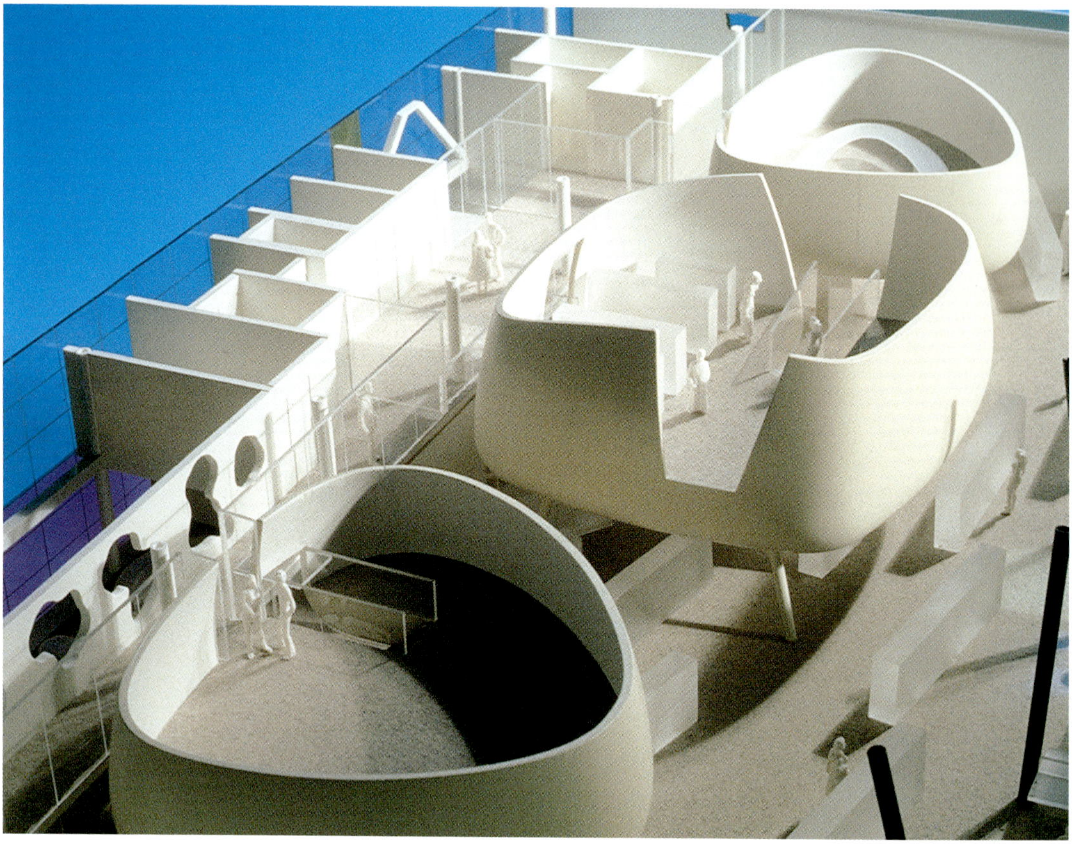

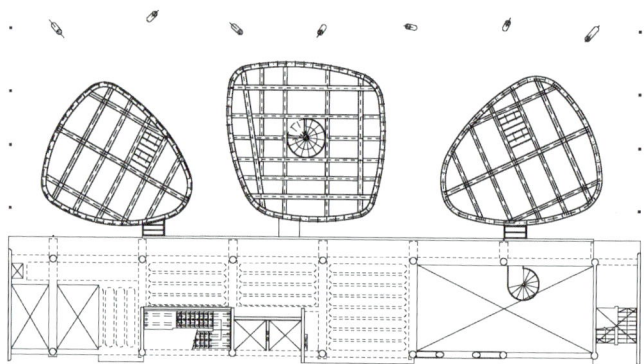

Three pods in the library The conceptual idea behind this library was to respond to the multiethnic character of the neighbourhood it serves. This was achieved by organising a series of distinct areas within a single big space: an Afro-Caribbean study centre, a children's room and a meeting room. Alsop's idea was to give these spaces the character of a floating object, and because the Afro-Caribbean centre was one of the spaces, he came to us with a pot — literally, an African pot — and said "I want something like this."

The pods allowed us to start thinking early on about how such forms are drawn, designed and made. The knowledge and redefinition of the disciplines involved have proved invaluable, as we have developed such geometries on larger scales.

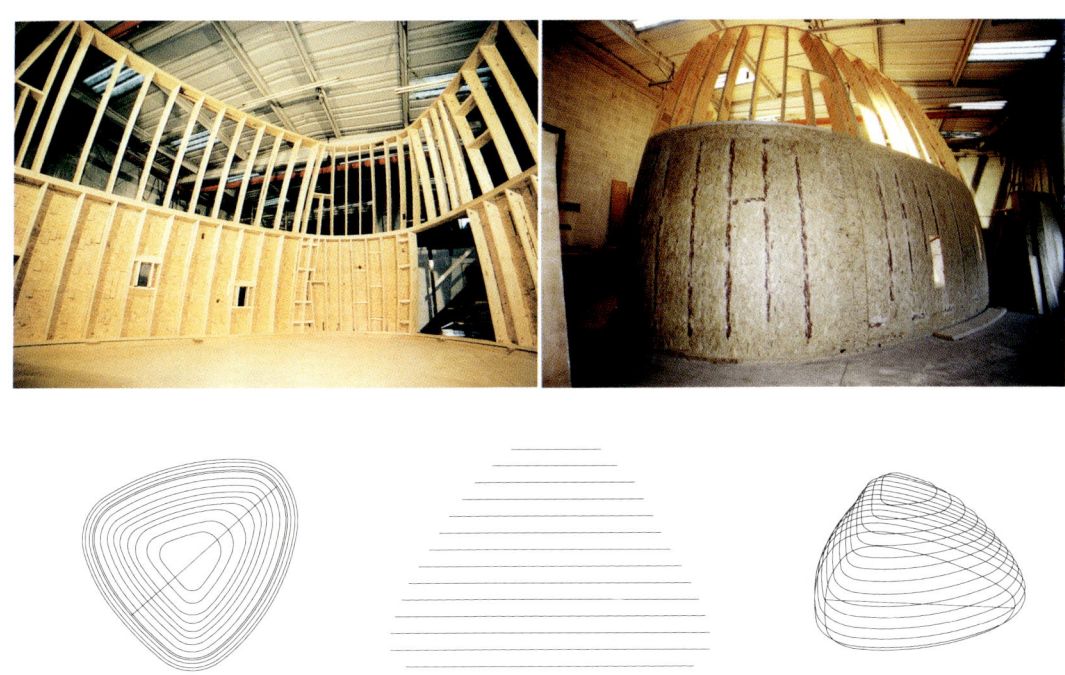

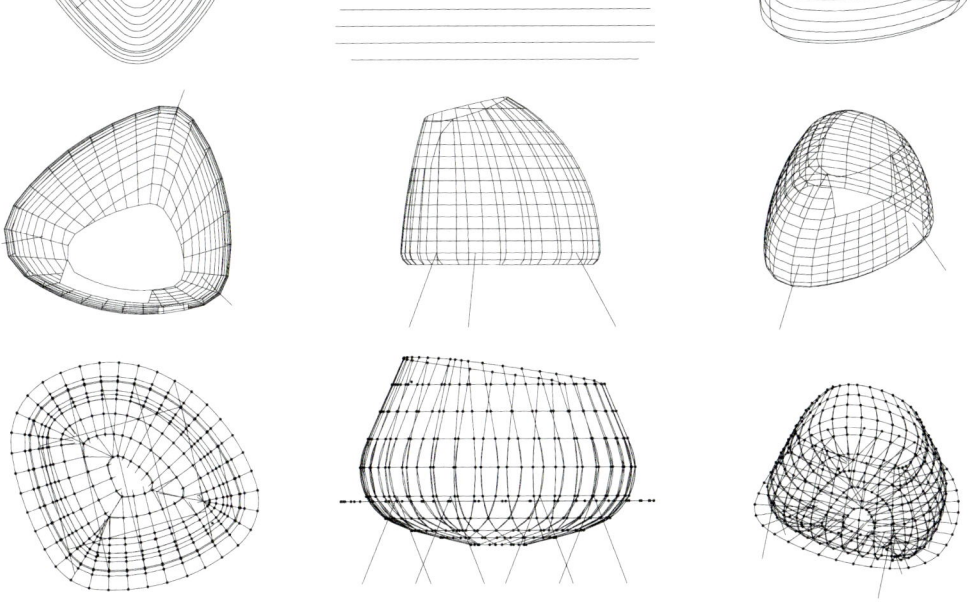

The pod's geometry

Our first challenge was figuring out how to draw this object. Even as recently as then, Microstation could not unfold the complex geometries of such forms to conventional plans and sections. Once we managed to draw it, the problem was then how to make it. The budget was limited, so we knew our solution had to be repetitive and from known technologies. After exploring ferrocement, advanced composite plastics and sprayed concrete, we settled on using microlam (or LVL, laminated veneer lumber). This is a very strong, manmade timber fiber produced in flat sheets, that the fabricator cut into 'hockey-stick' shapes. These pieces (this was during the early days, before architects like Greg Lynn started automating such forms) could then be glued together to make a wall joist of any shape.

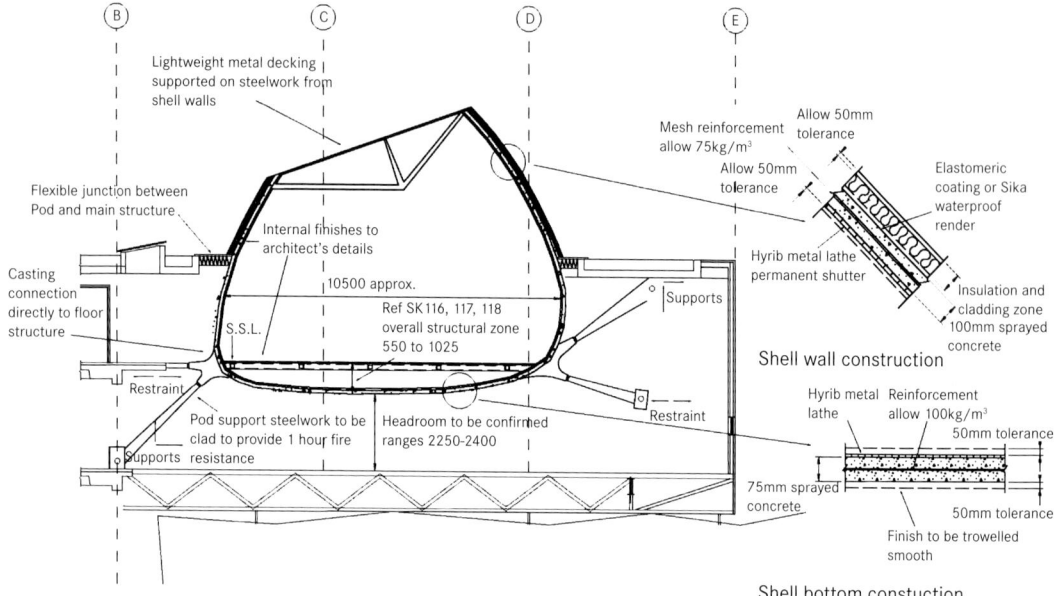

Pod's section

By creating these hockey stick elements, the forming of doubly curved timber pods becomes a non-issue as from a flat sheet of the material, you can cut any shape you want.

We then put a skin on both the inside and the outside, to encourage stress skin action. Originally, the architect wanted to clad the exterior in leather (he had this curious notion of using cow leather as a general cladding material to relieve the disposal of leather). This was not practical, so what you see on these pods is a very thin 3mm layer of the same timber in the form of tiles that are stapled to the frame to make it look like leather. It is a beautiful object that does actually look like a pot — the pot that he first gave us.

ACTAR: For AKT this was a classic rationalization problem: an architect wants to build something and you have to figure out how to do it.

We had to figure out the most economic way to do it without inventing a new material, or coming up with a one-off solution. We needed to produce something that could be easily made with good craftsmen and applied to other projects. Precedents from boat making and conventional timber floors inspired this approach. At the time, producing fabrication drawings for this type of structure was the biggest challenge.

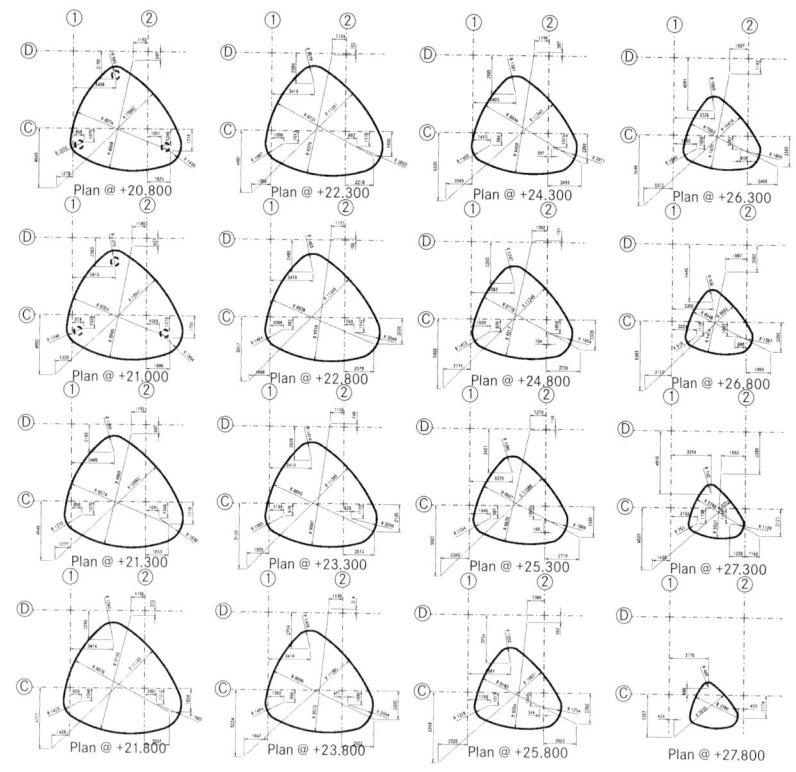

Plan @ +20.800 Plan @ +22.300 Plan @ +24.300 Plan @ +26.300

Plan @ +21.000 Plan @ +22.800 Plan @ +24.800 Plan @ +26.800

Plan @ +21.300 Plan @ +23.300 Plan @ +25.300 Plan @ +27.300

Plan @ +21.800 Plan @ +23.800 Plan @ +25.800 Plan @ +27.800

Pod A. Setting out

Pod B. Setting out

Plan @ +20.950 Plan @ +21.450 Plan @ +22.950

Plan @ +21.150 Plan @ +21.950 Plan @ +23.450

Plan @ +21.312 Plan @ +22.450

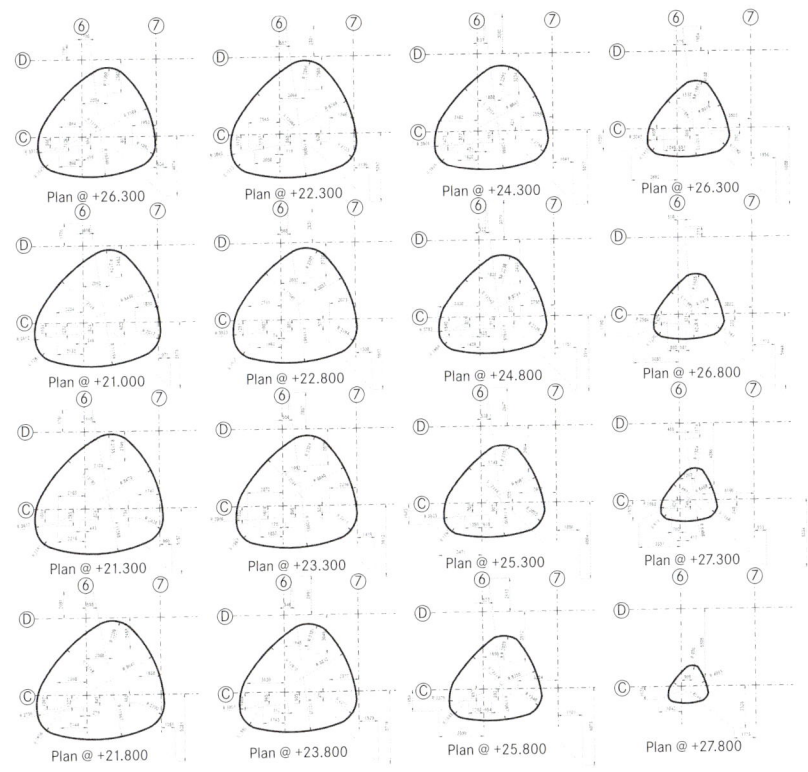

Plan @ +26.300 Plan @ +22.300 Plan @ +24.300 Plan @ +26.300

Plan @ +21.000 Plan @ +22.800 Plan @ +24.800 Plan @ +26.800

Plan @ +21.300 Plan @ +23.300 Plan @ +25.300 Plan @ +27.300

Plan @ +21.800 Plan @ +23.800 Plan @ +25.800 Plan @ +27.800

Pod C. Setting out

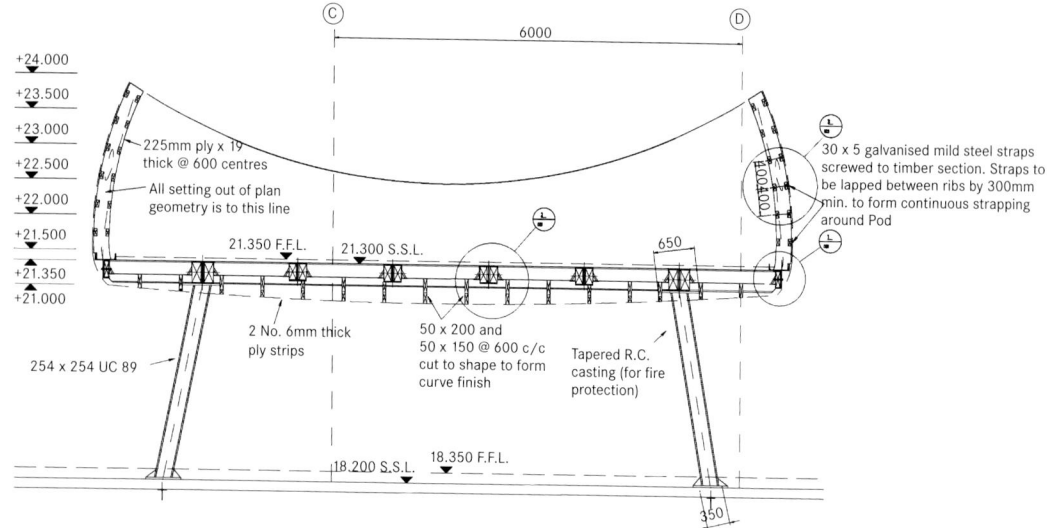

Pod Section

Pod plan at levels 4 and 5

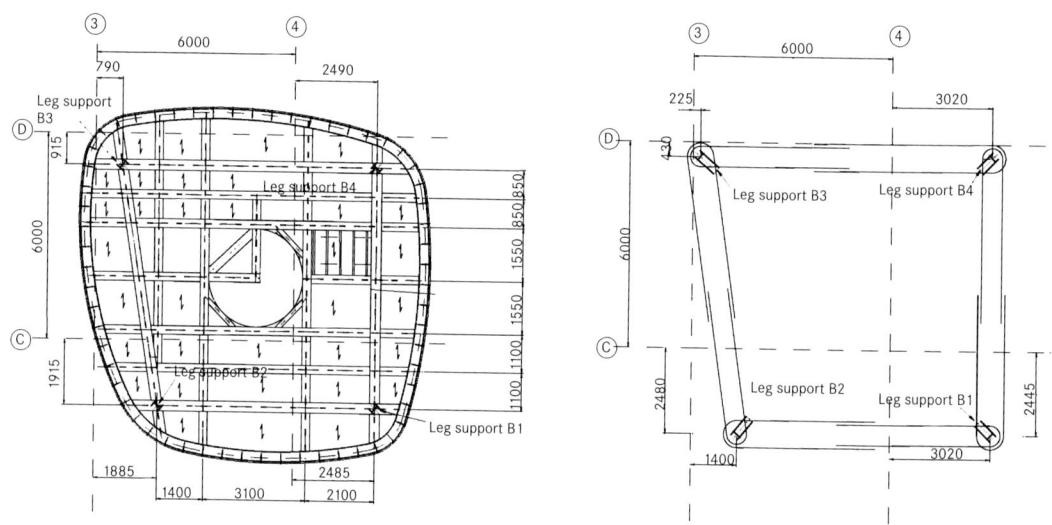

Will Alsop's vision for this 10,500 square-meter research and teaching complex for Queen Mary College's medical and dentistry departments combines an overall rational structure with specific, irregular interventions. Up to 400 researchers can work simultaneously in the laboratory space on the single basement floor. Above, raised on columns, "hovers" a large single volume with numerous irregular insertions, a bulbous lecture theatre, closed offices and open floating meeting spaces. Alongside these insertions a 'plant wall' concentrates all the service and maintenance requirements in a separate structure that plugs into the laboratories as necessary. AKT's structural concept standardizes as much of the structure as possible, and results in one system for the overall structure and a second one for the "plant wall."

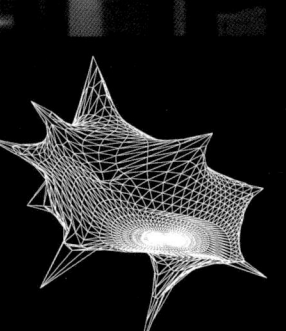

PODS
INSTITUTE OF CELL AND MOLECULAR SCIENCE

ALSOP ARCHITECTS
LONDON, 2006
CLIENT: QUEEN MARY COLLEGE, UNIVERSITY OF LONDON

After the success of Peckham, where Alsop rethought library space, we teamed up again in rethinking the definition of a laboratory style. We secured this through a competition amongst engineers and with the architect started rethinking this type with the simple and common perception that scientists like to be secretive so, architecturally, you hide them. The architect's concept was (as in the library) to make an important semi public building, and to open up the space so that the scientists who either want to merge and collaborate or work in isolation will be attracted to come and work here. Today many universities are competing to have world-class laboratories and if you don't have a 'designed' building, you don't attract the best scientists. Therefore, this project needed to be both 'iconic' and 'emblematic' for Queen Mary and Westfield College.

In laboratories, service engineers have to run service and plant pipes, which always dominate the architectural and structural intent. In response, we moved the plant space out and made a special 'plant wall' that connects underground to the main building. We created a public square in between the two buildings, which has provided a well used central public space, like the space under Peckham Library.

For the laboratory spaces the architect wanted to continue exploring pod construction. He gave us a physical model, each pod having a label such as 'Spikey' or 'Brain', and we were given the freedom to experiment with how we might make them; today, this is one of the only ways in which one can innovate. Coincidentally, one of them was made from a material we had always wanted to try; advanced composite plastic.

It looks like a red blood cell?

But it's orange really, brightly coloured. So, we continued to play with the pods. At some point one person in the office actually collected fifteen of them in a little booklet. These objects attract a lot of interest. People love the fun aspect of this building and scientists have been very complimentary about working in it.

And each pod has a different construction system?

Yes, Spikey is PVC fabric (a commonly used woven engineering fabric). We have always been fans of trying to use plastic in buildings, but hardly ever get the opportunity. Another pod is timber, like at Peckham. We went through all sorts of manufacturing and assembly options while developing them. So as a continuation of the enquiry, the work from Peckham was dramatically taken forward.

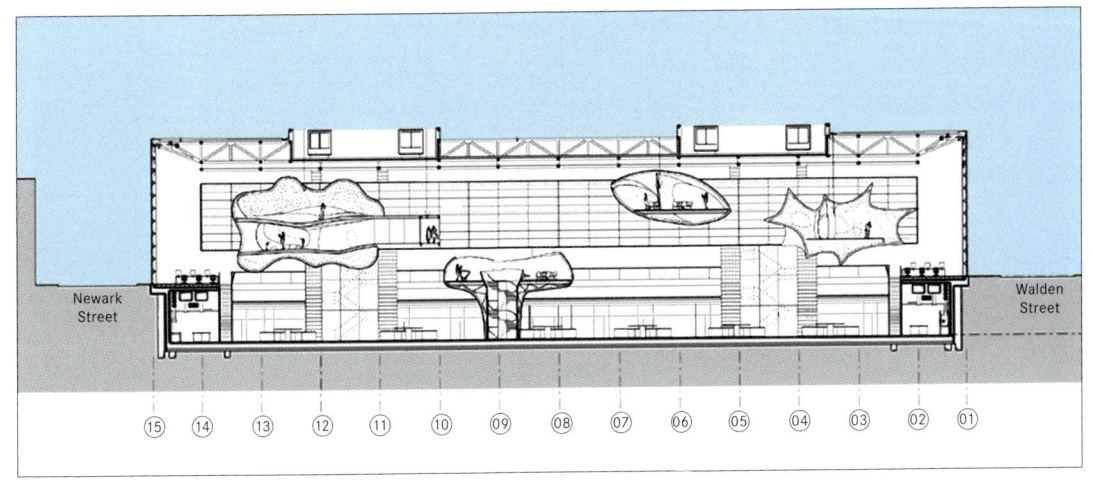

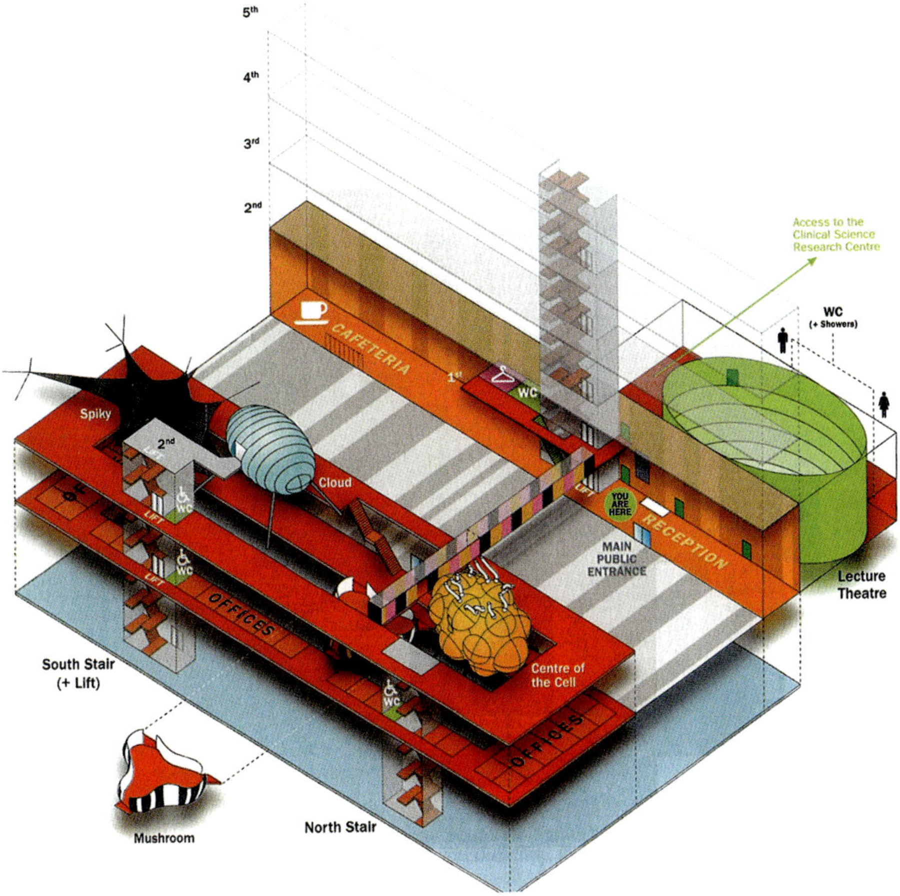

Section and isometric view of the building

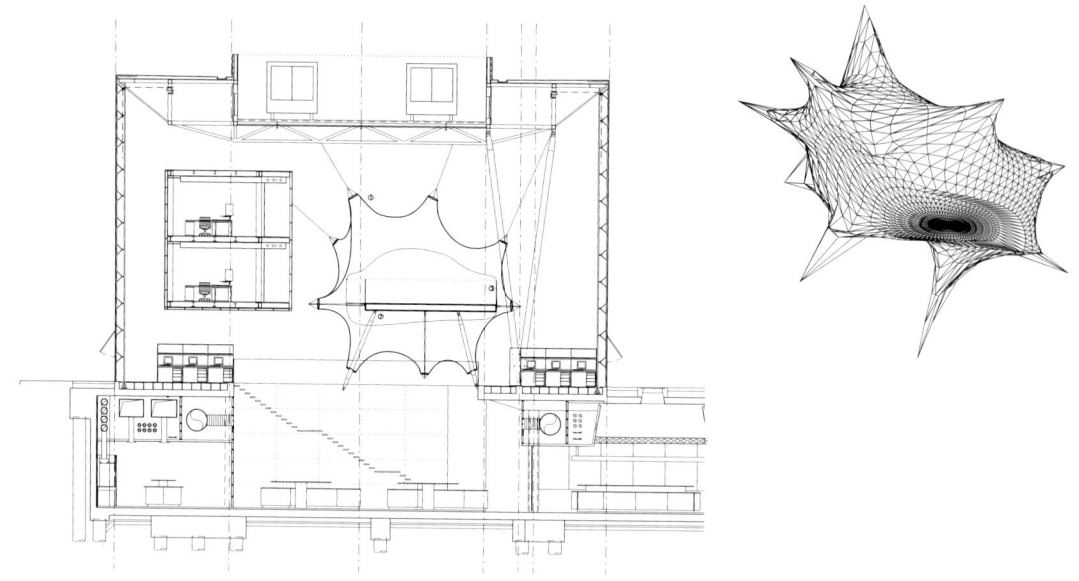

Spiky: Section in the building

Nodal forces

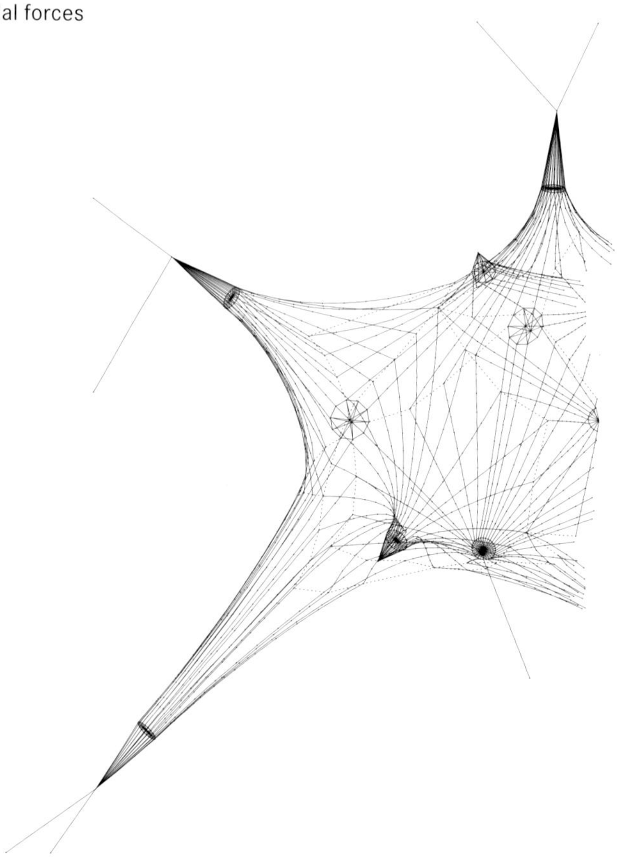

General assembly - conic 2

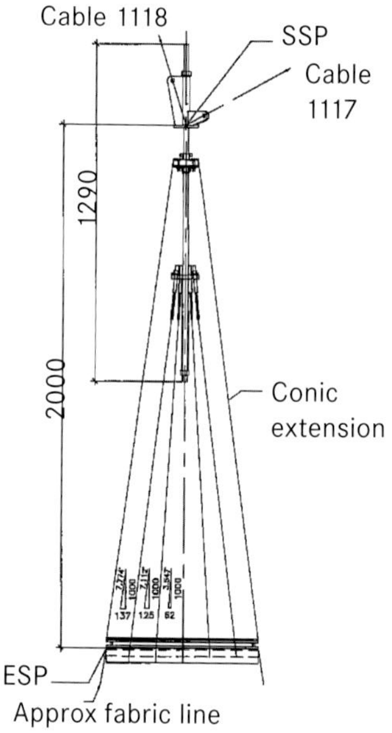

Cable 1118

SSP

Cable 1117

1290

2000

Conic
extension

ESP

Approx fabric line

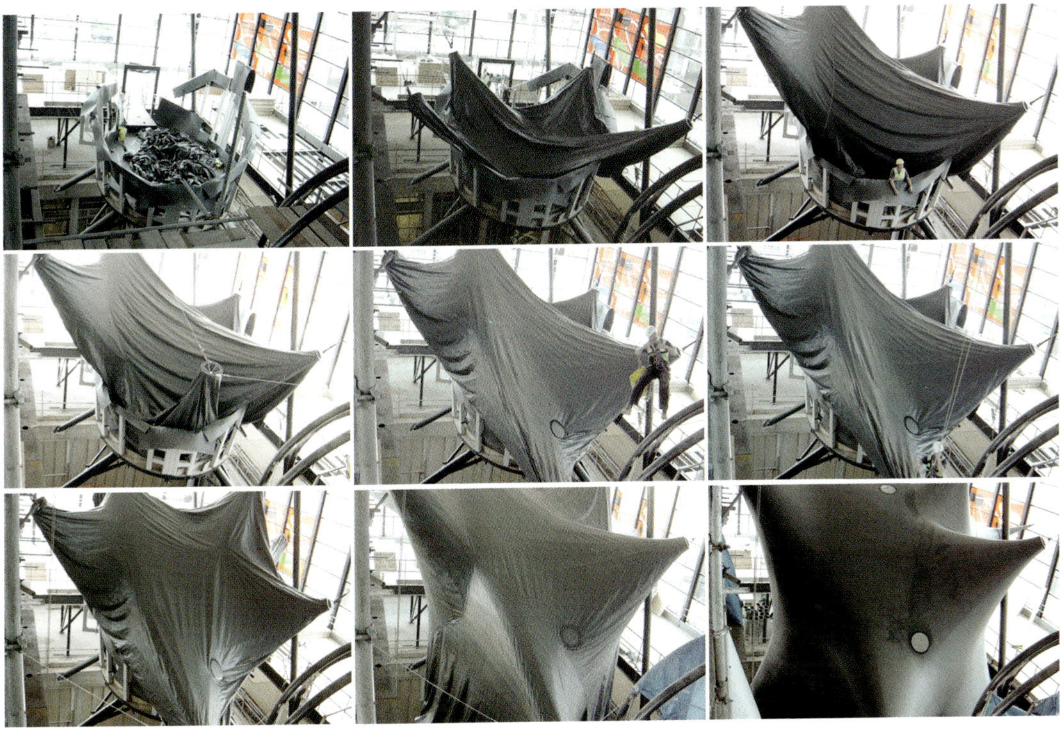

Spiky in construction

Finished

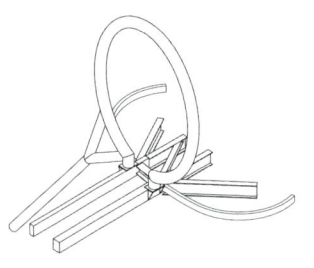 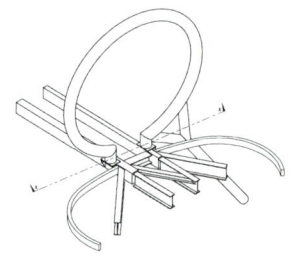 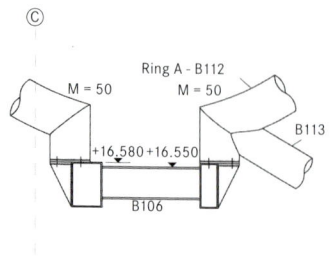

Isometric views on Ring A

Section detail

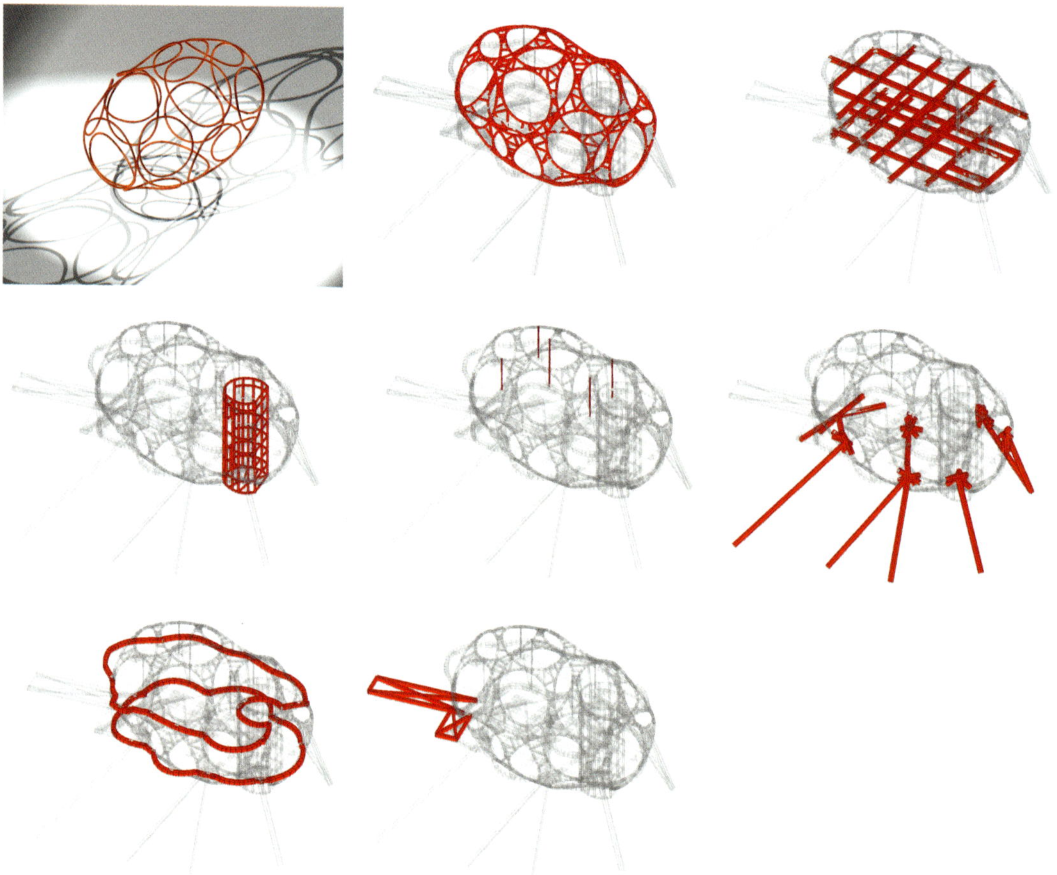

Ring A - B112
M = 50   M = 50
B113
+16.580 +16.550
B106

Multiframe analysis of the Centre of the Cell

Construction process of the Centre of the Cell

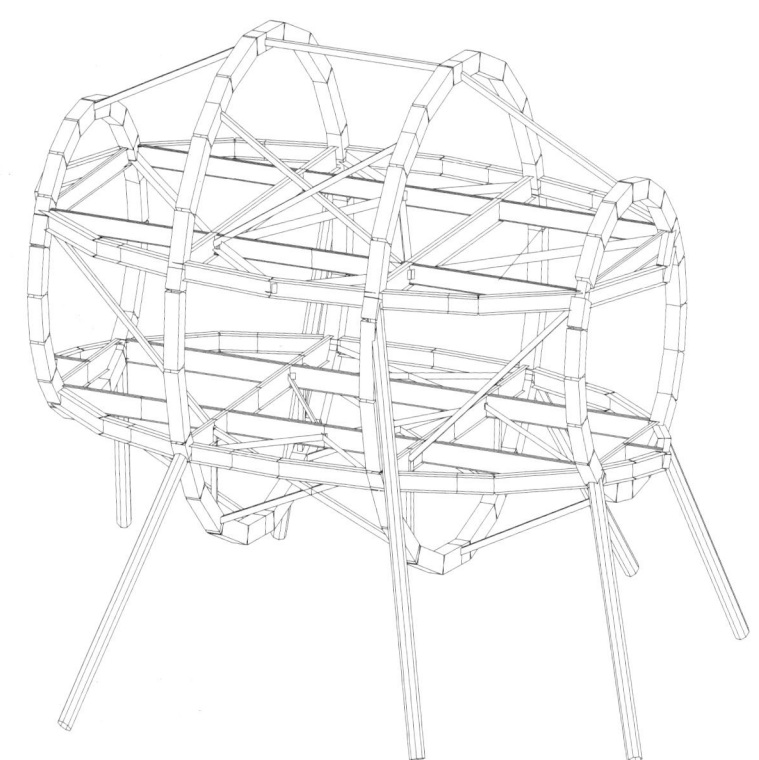

Cloud: Structure

Cloud: Plan

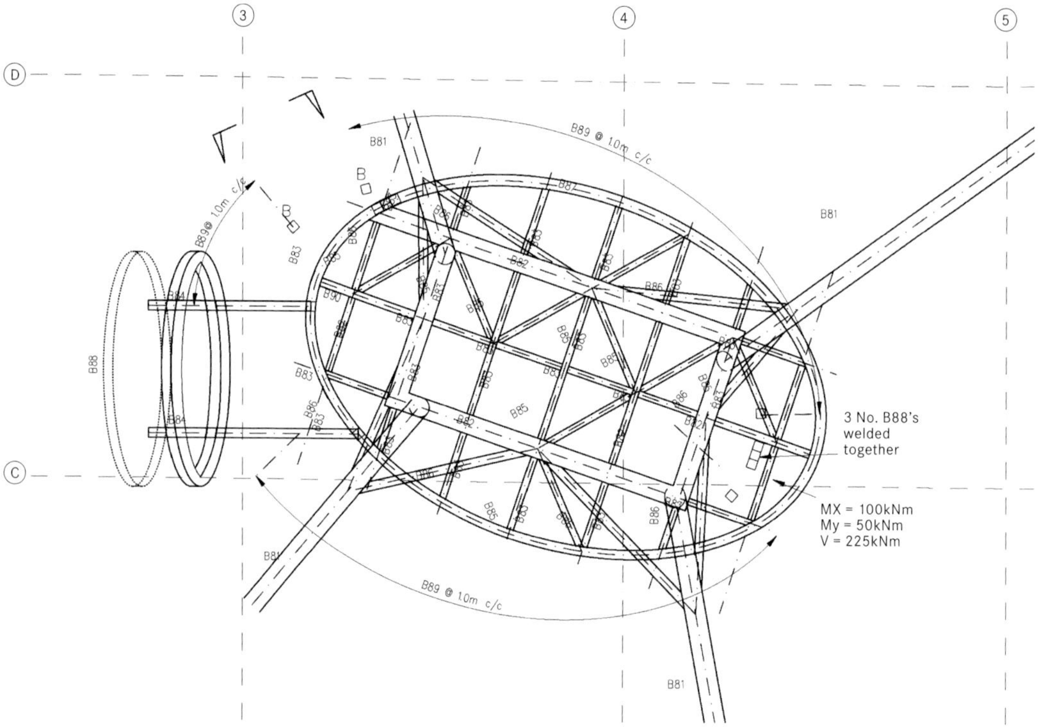

AKT's collaboration with Foreign Office Architects began with Belgo, a Belgian restaurant built as a conversion of a disused cinema in Ladbroke Grove, London. The concept of the building is based on a series of vaulted shell surfaces that merge wall and roof, bringing daylight into the space and loosely evoking the forms of mussel shells. By inclining the vaults, the roof of the dining room appears from one end to be a continuous surface with concealed light sources, while from the other end it appears as a series of thin, delicately poised ribbons. AKT developed a simple structural solution for these complex roof forms by customizing standard symmetrical steel arches with conventional joists, plywood, and a standing seam roof construction. All structural loads are resolved into the vaults, a solution for providing supports on a restricted site with minimal disruption to surrounding buildings.

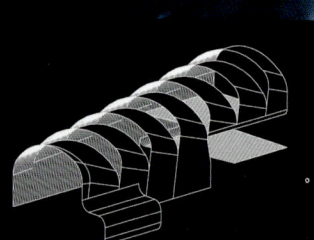

ROOF ARCHES
BELGO ZUID

FOREIGN OFFICE ARCHITECTS
LONDON, 1999
CLIENT: BELGO

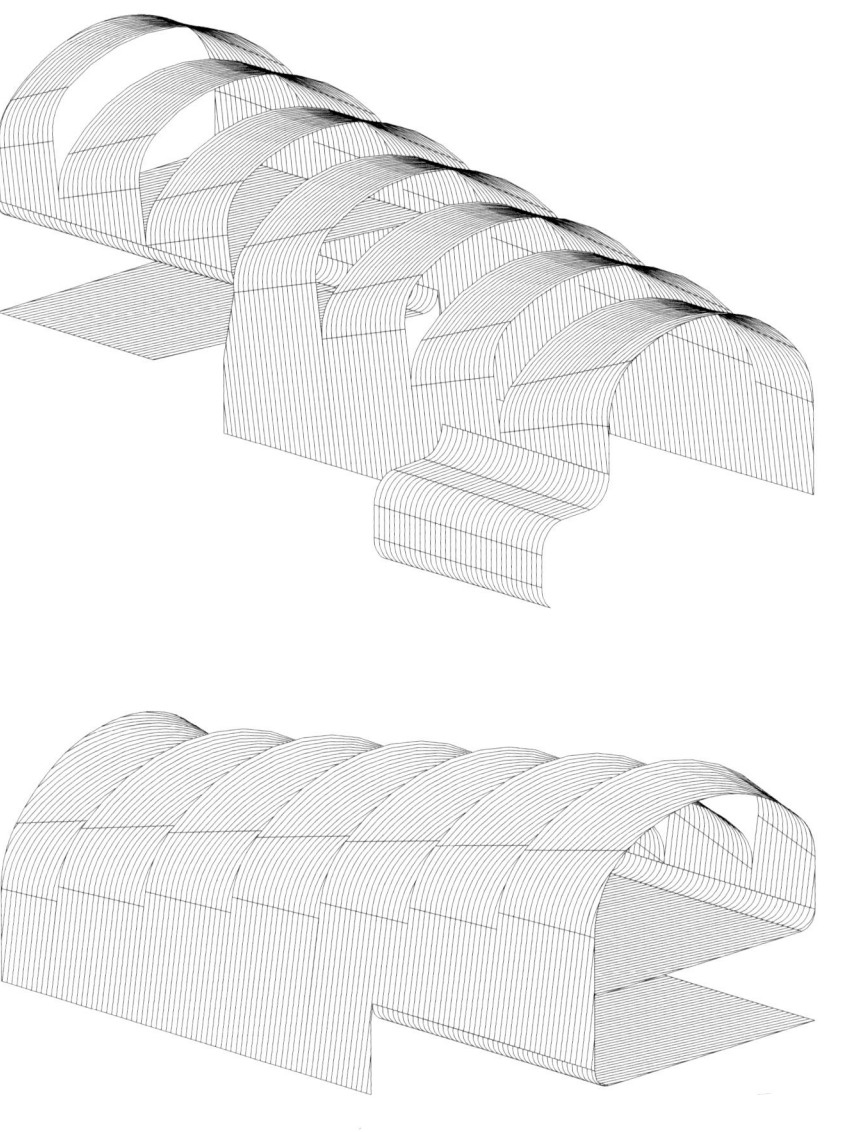

FOA's initial concept drawing

For this project, the client and FOA invited three engineers for an interview. We had to comment on how to deliver the initial concept themed from 'mussels' and seafood. We had already visited the site, so we knew that there were many contextual constraints we would have to work through.

For instance, it was locked in between numerous existing buildings with little access, so even walking onto the site was difficult. The challenge lay not with the production of their design proposal, but with simple logistics, such as getting a construction plant into that space. Inserting the conceptual idea onto existing structures and a constrained site 'bred' the structural logics. Peripheral issues such as avoiding noise during construction become more important parts of the design in such situations.

FOA's research and enquiry into surface structures where wall becomes floor became a key driver, relegating the structure into simply delivering this shape within the constraints.

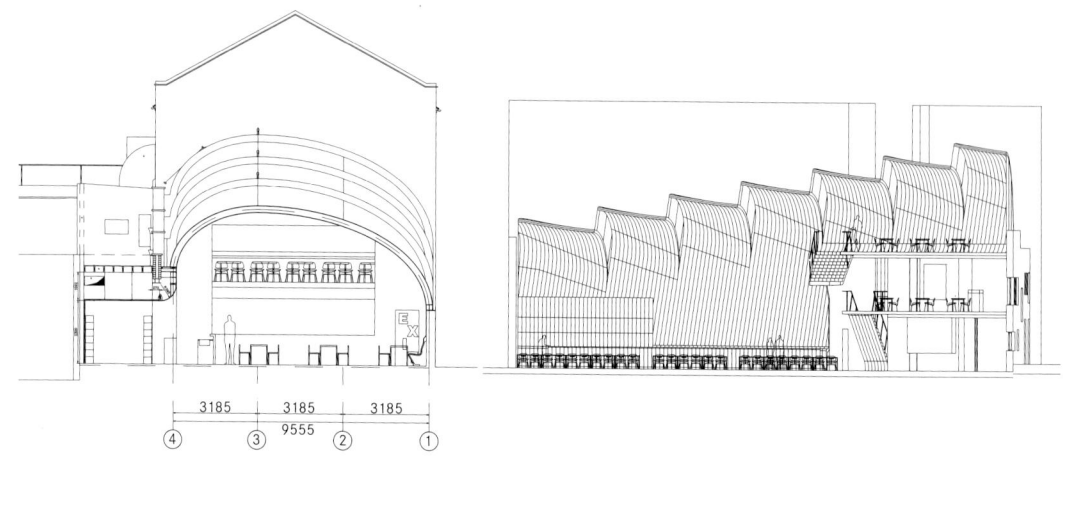

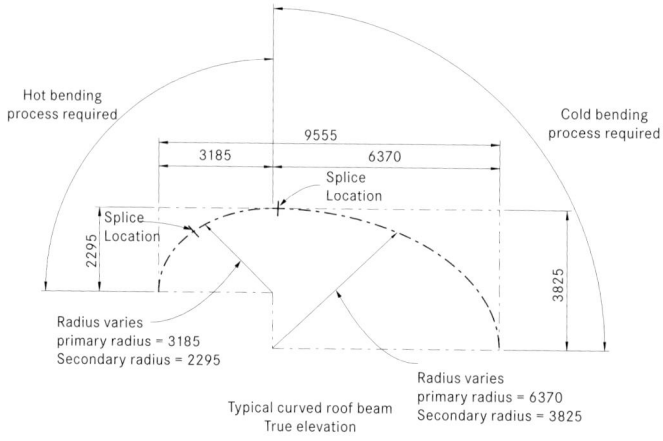

Hot bending process required

Cold bending process required

9555

3185

6370

Splice Location

Splice Location

2295

Radius varies
primary radius = 3185
Secondary radius = 2295

3825

Typical curved roof beam
True elevation

Radius varies
primary radius = 6370
Secondary radius = 3825

Typical setting out drawings

Concurrently, we were working on Peckham Library, and therefore had learnt a lot about contemporary use of timber and about how to bend it — knowledge which helped make this project. Stressed skin plywood 'shells' spanning onto curved steel section was the optimum solution at Belgo. For the steel, it was generally felt that a single-radius curve is cheaper than one that changes several times. However, with advances in technology, this was no longer the case. Once you curve something you pay a premium, but following the curve correctly into a working stress pattern (compound curve) costs the same amount. We wanted to make shells completely out of timber, which we could take to site as plywood sheets, and then bend in place to overcome the cranage and site access constraints.

The principle developed on the understanding that when a continuous shell is "torn" (in this case, to generate the stepped roof) weakness is created at the 'tear' as continuity of the overall shell is compromised. At this tear, the timber must be reinforced by a steel rib and it becomes a composite structure of steel and wood. In structural terms, it's a sort of hybrid between a shell and a conventional trussed roof that some refer to as semi monocoque.

The limited use of steel elements helped to keep the project affordable, reducing the number of large pieces to transport and erect to only six or eight curved pieces. These included a splice at the bottom of each vertical where it connected to the base; the success lies in the production of the complex curve with the right combination of offsite fabrication and on-site construction to install the structural shells in eight hours. The bending of all the plywood sheeting was on-site (plywood in 3mm and 6mm thickness allow for manually bending) allowing 'low order' craft to deliver the perceived contour shape. The simplicity of this double-layer of 'plywood' also permitted simple fastening of the termed stainless steel to the outside and the 'patterned' timber finish to the inside to allow the architect to create the single surface effect internally. The exterior was simply coated with rubber and covered with stainless-steel rain screen.

That was our first project with FOA. I think they appreciate the fact that we connected with the industry quickly, and proved that things could be done within very short deadlines, even though the client didn't think it would be possible. The project allowed AKT to seed the FOA work into the design process, which has then grown with more collaborations.

The seven-story, 15,000 square-metre headquarters for fashion retailer Monsoon Accessorize fulfills the clients' need for an open plan to accommodate offices and design studios. Moving the support to the exterior façade, AKT designed a highly efficient structural system — a lattice of inclined columns that channel lateral, as well as vertical, loads. That structural efficiency helps floor plate efficiency because it reduces the role of the cores in providing lateral stability. As it rises up the dimensions of the concrete 'net' diminish and the design terminates in the steel support for a saw tooth roof. The floors surround a central atrium, which divides each level into two eighteen by forty-five metre zones. Plans for an additional four buildings on the site will eventually double the accommodation.

NET STRUCTURE
MONSOON VILLAGE

ALLFORD HALL MONAGHAN MORRIS
NOTTING HILL, LONDON, 2007
CLIENT: HOUGUE LIMITED

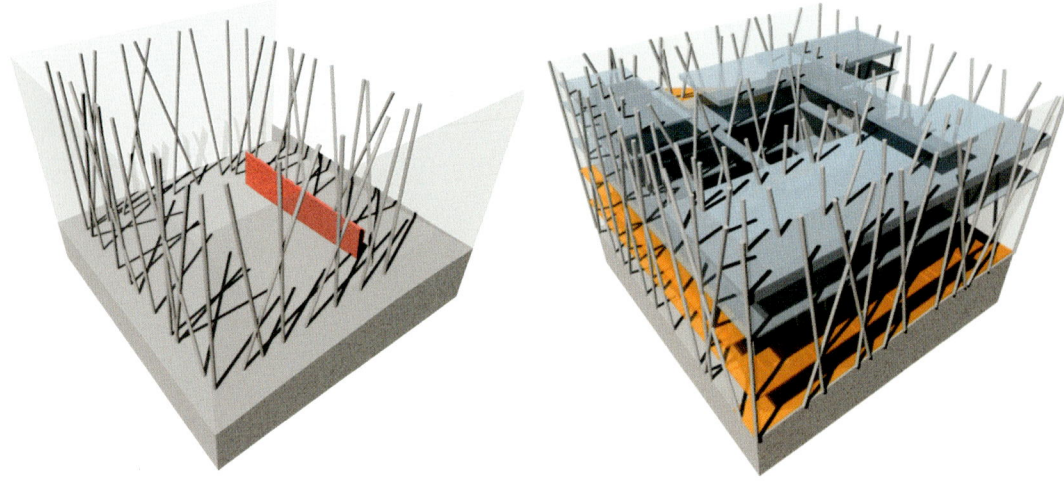

The client had charged AHMM to a unique headquarters building with a legible identity but one that could also be rented as a generic office space. AHMM's architectural approach is usually very pragmatic, not prescriptive, and demands a very similar approach from the engineers. At Monsoon they make fabric, so we proposed a branching, net structure as a way to convey an interesting identity to the façades.

Using CATIA — which we had recently incorporated in the office — we prepared a series of parametric models that illustrated the simplicity of a net structure that did away with the typical columns. The net is reminiscent of a fabric covering, and can be paired with a regular glass façade. It's a low-tech and low-cost idea and, following our experience at Peckham Library and the Phaeno Science Centre, we were confident that it could be built in reinforced concrete without any problems.

Did your structural options then actually get the architects interested into this type of complex structure?

Yes, and this in a way reflects the dynamics of our own office. We allow ourselves total freedom to do almost anything that's good engineering, without letting it completely blow out in all directions. Control is maintained by being very critical about each other's passions while encouraging each other's strengths. This is what defines a common envelope around our approach.

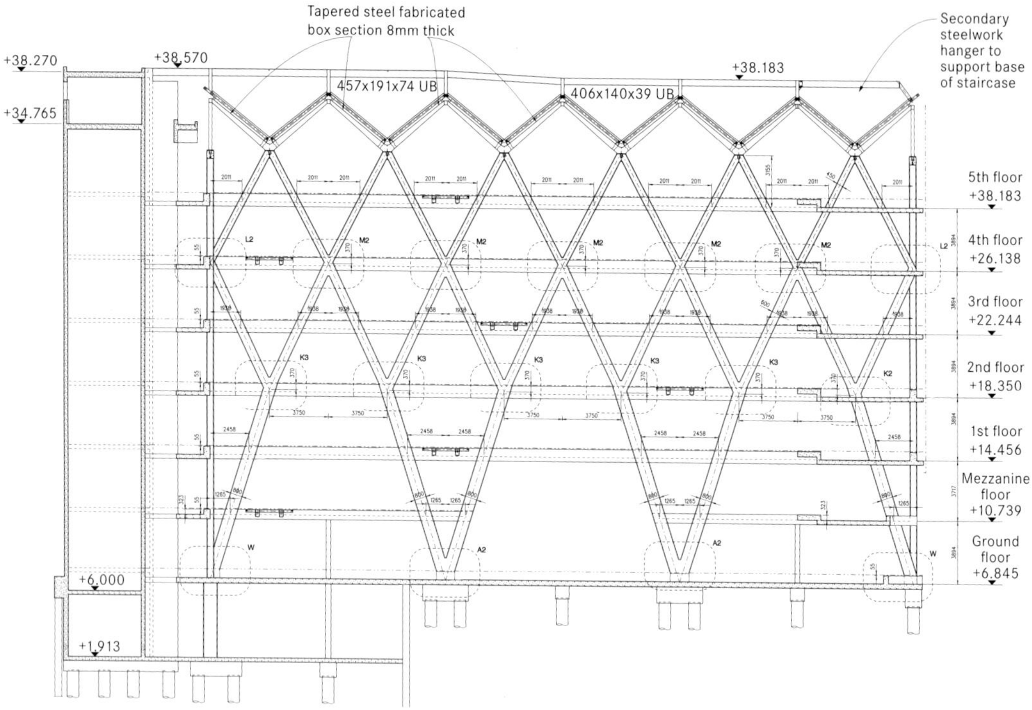

Tapered steel fabricated
box section 8mm thick

Secondary
steelwork
hanger to
support base
of staircase

+38.270

+38.570

+38.183

+34.765

457x191x74 UB

406x140x39 UB

5th floor
+38.183

L2   M2   M2   M2   M2   L2

4th floor
+26.138

3rd floor
+22.244

K3   K3   K3   K3   K2

2nd floor
+18.350

1st floor
+14.456

Mezzanine
floor
+10.739

W   A2   A2   W

Ground
floor
+6.845

+6.000

+1.913

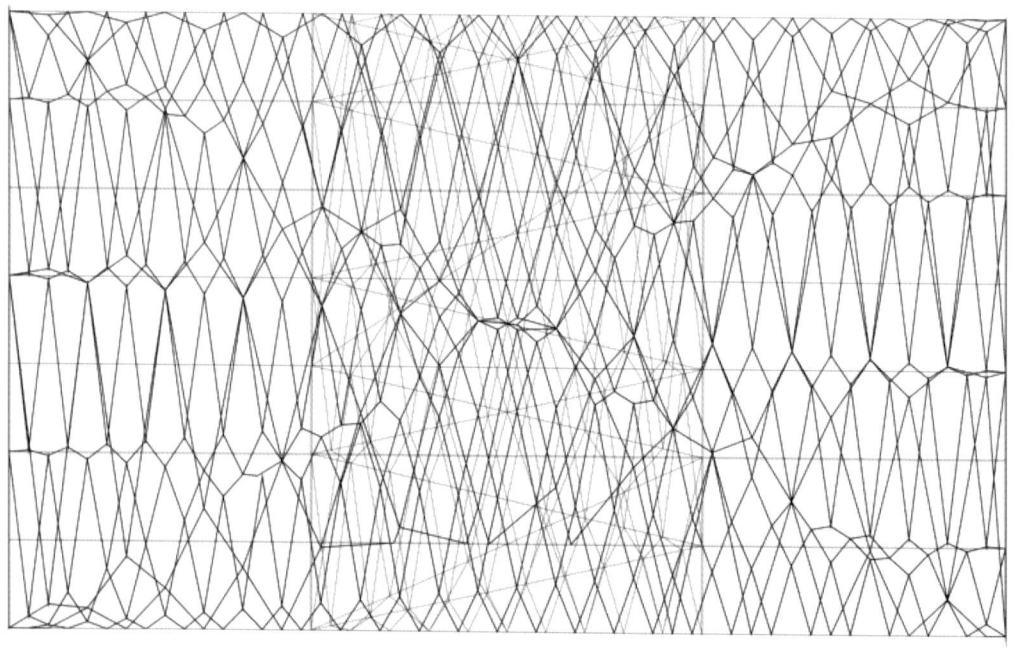

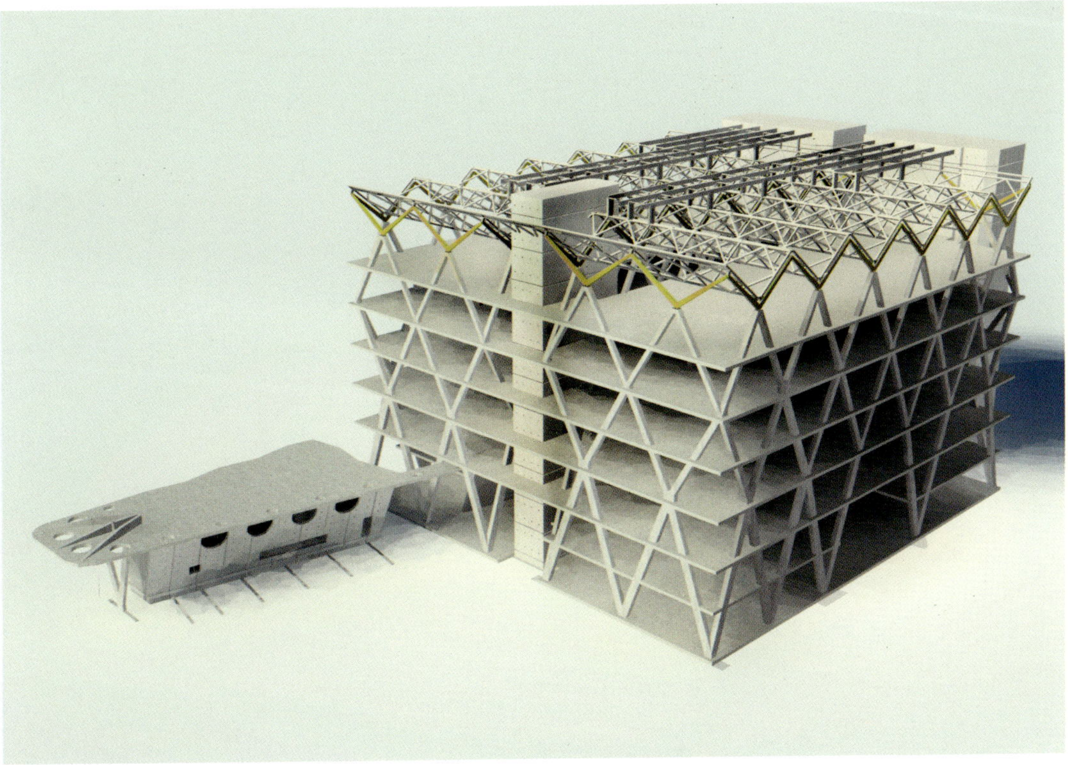

Construction images

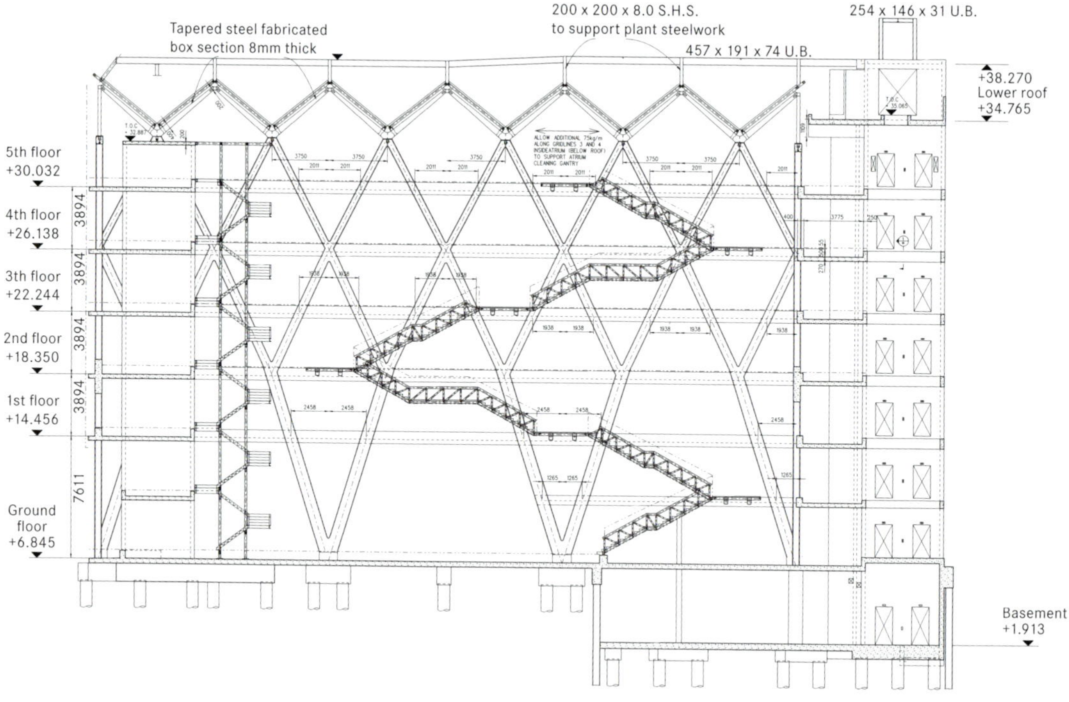

Tapered steel fabricated
box section 8mm thick

200 x 200 x 8.0 S.H.S.
to support plant steelwork

254 x 146 x 31 U.B.

457 x 191 x 74 U.B.

+38.270
Lower roof
+34.765

5th floor
+30.032

4th floor
+26.138

3th floor
+22.244

2nd floor
+18.350

1st floor
+14.456

Ground
floor
+6.845

Basement
+1.913

3894

3894

3894

3894

7611

Elizabeth House, a competition entry designed with Foreign Office Architects for a tower above Waterloo Station in central London, was one of our first opportunities to consider the relationship between structure and architecture in a tall building. The key was to find a logical way of springing the tower from the complicated network of tunnels and structures below ground. Once the foundation positions were understood, it was tempting to extrude identical floor plates upwards, but FOA chose to develop a new prototype for tall buildings. This led to the idea of trapezoidal floor plans, where each floor plate varies slightly, creating a continually changing profile. This not only adds visual interest to the skyline, but also means that each floor uniquely offers a slightly different view, acknowledging existing sightlines and monuments within the urban fabric.

With this concept, the logical location for the primary structure is on the outside perimeter, where it forms a tube that acts in conjunction with the central core. A conventional extruded tower would repeat its form on every floor. This FOA proposal repeats every sixteen floors, and the arrangement of the plan helps to establish a triangulated external truss on the facade. This itself opens the range of possibilities for expression, and FOA carefully added to them by introducing a pattern of opacity and translucence in the windows, whose hexagonal shape fits into the structural form. Though it would have been possible to take out some of the structure on the upper levels, we decided to keep the façade homogeneous to convey the fact that this design could be scaled up or down, reinforcing its 'prototypical' nature.

INTEGRATION OF DESIGN PROCESS
ELIZABETH HOUSE

FOREIGN OFFICE ARCHITECTS
WATERLOO, LONDON, 2005
CLIENT: P&O ESTATES

We needed to find a pattern that would work for the structure and the façade throughout the height. The first renderings seemed to indicate that it was possible to develop the geometry from a regular floor plate with slight adjustments to the floor plan. Although some triangular spaces are not considered commercially viable, they provided unusual spaces with specific views.

We developed a system to make this form in a simple and repetitive manner. By dividing the floor plate equally, you can simplify its construction and you create points of support for the vertical structure on the façade. These points are extruded down in a diagonal pattern, hence acting as diagonal columns whilst also assisting the 'soft tube' principle for lateral stability.

This produced an external primary mega-structure which provides a significant stiffness laterally and, with the closed module, ideal support for the façade. This facilitated a series of options for making the tower more transparent as it ascends. In the end, we used a very integrated solution that involved making the floor plate conventionally out of composite constructions and supporting it by a diagrid instead of vertical columns. That diagrid was then coordinated with the façade design to study its opacity, and the analysis showed that the weight of steel was similar to that of a conventional tower.

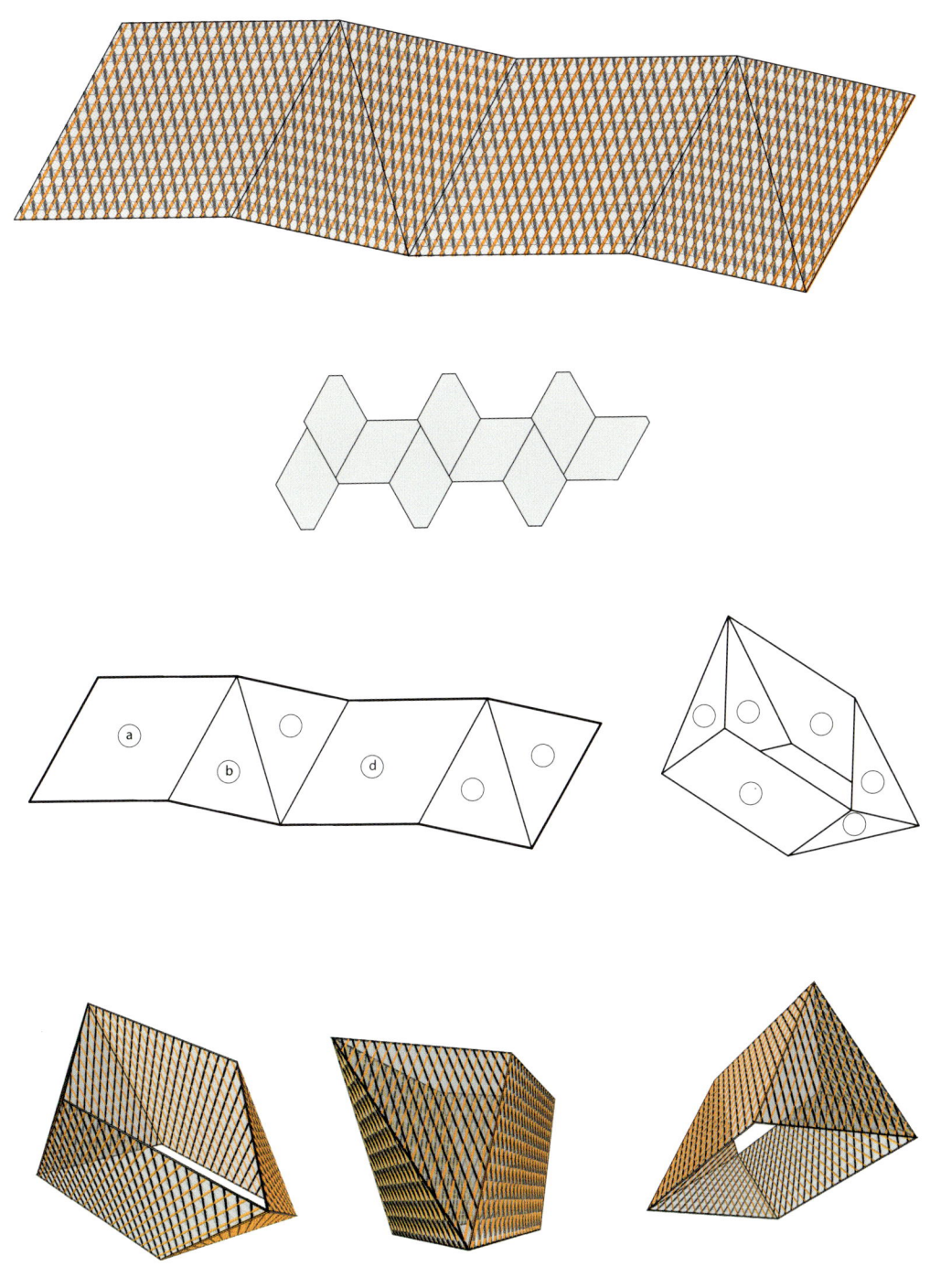

Unfolding a module: Hexagon/Herringbone pattern

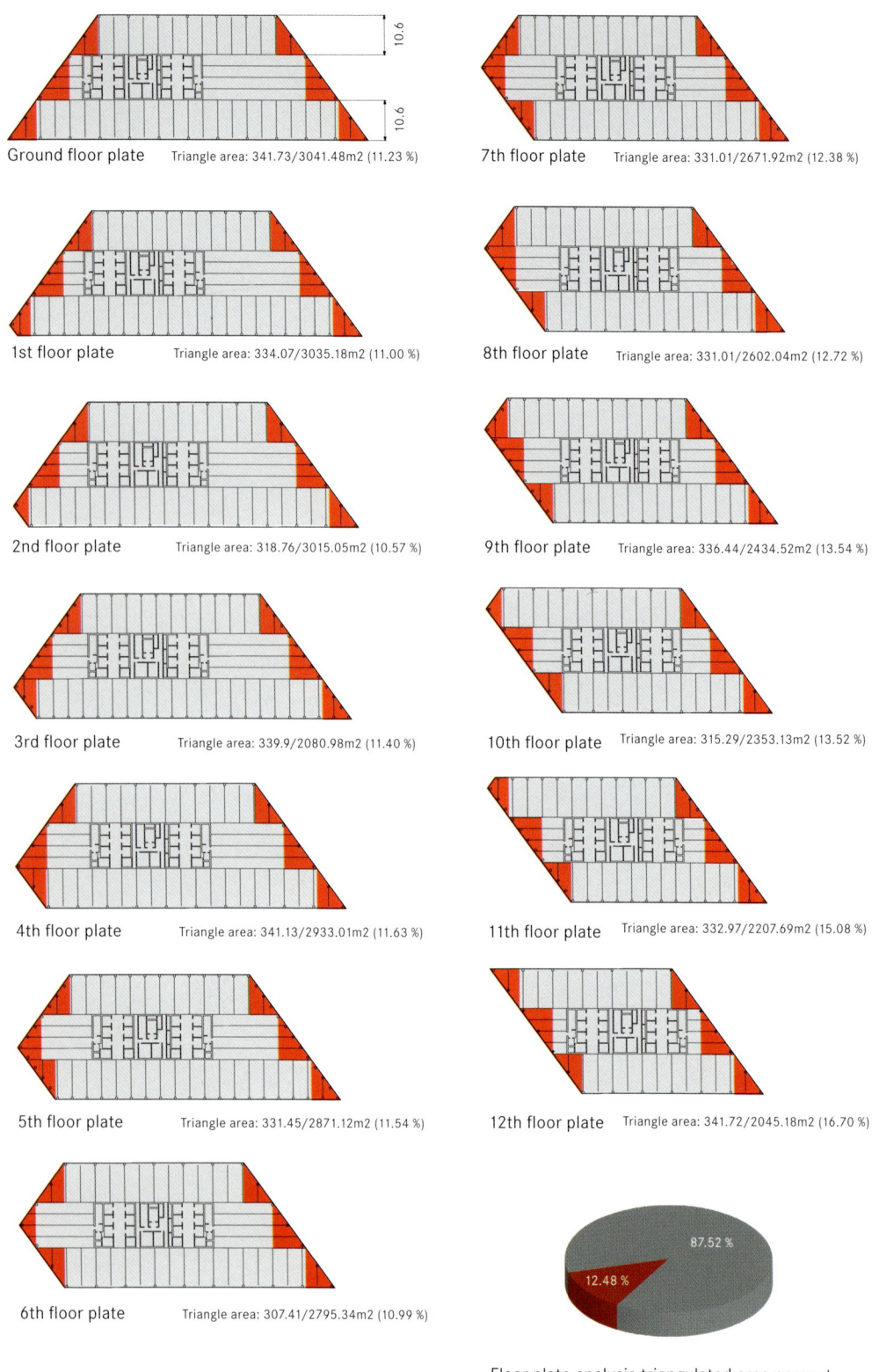

Ground floor plate   Triangle area: 341.73/3041.48m2 (11.23 %)

10.6

10.6

7th floor plate   Triangle area: 331.01/2671.92m2 (12.38 %)

1st floor plate   Triangle area: 334.07/3035.18m2 (11.00 %)

8th floor plate   Triangle area: 331.01/2602.04m2 (12.72 %)

2nd floor plate   Triangle area: 318.76/3015.05m2 (10.57 %)

9th floor plate   Triangle area: 336.44/2434.52m2 (13.54 %)

3rd floor plate   Triangle area: 339.9/2080.98m2 (11.40 %)

10th floor plate   Triangle area: 315.29/2353.13m2 (13.52 %)

4th floor plate   Triangle area: 341.13/2933.01m2 (11.63 %)

11th floor plate   Triangle area: 332.97/2207.69m2 (15.08 %)

5th floor plate   Triangle area: 331.45/2871.12m2 (11.54 %)

12th floor plate   Triangle area: 341.72/2045.18m2 (16.70 %)

6th floor plate   Triangle area: 307.41/2795.34m2 (10.99 %)

87.52 %

12.48 %

Floor plate analysis triangulated area percent
from the ground floor to the twelfth floor

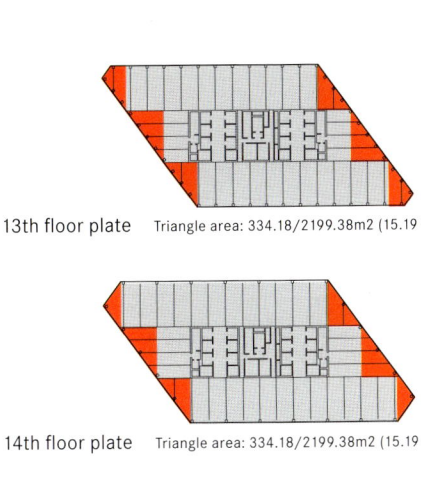

13th floor plate   Triangle area: 334.18/2199.38m2 (15.19 %)

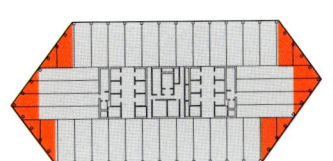

19th floor plate   Triangle area: 323.66/2530.71m2 (12.78 %)

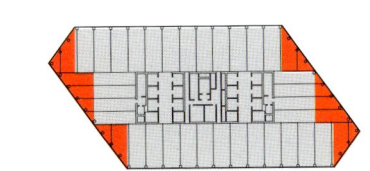

14th floor plate   Triangle area: 334.18/2199.38m2 (15.19 %)

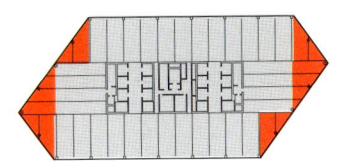

20th floor plate   Triangle area: 343.22/2487.99m2 (13.79 %)

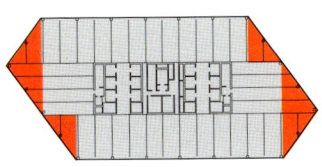

15th floor plate   Triangle area: 333.53/2421.38m2 (13.77 %)

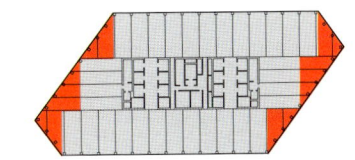

21st floor plate   Triangle area: 334/2418.79m2 (13.80 %)

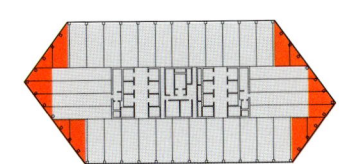

16th floor plate   Triangle area: 342.06/2490.75m2 (13.73 %)

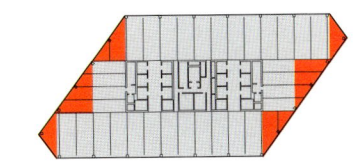

22nd floor plate   Triangle area: 297.02/2321.93m2 (12.79 %)

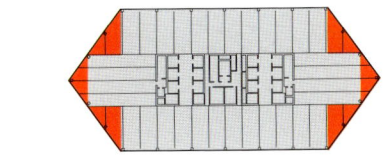

17th floor plate   Triangle area: 322.57/2532.09m2 (12.73 %)

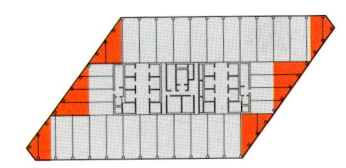

23th floor plate   Triangle area: 326.92/2197.38m2 (14.87 %)

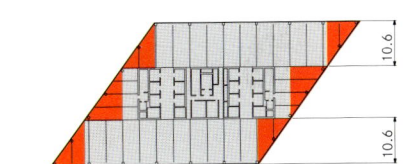

18th floor plate   Triangle area: 277.93/2545.41m2 (10.91 %)

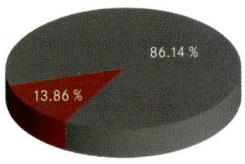

24th floor plate   Triangle area: 341.75/2045.18m2 (16.71 %)

Floor plate analysis triangulated area percent
from the thirteenth floor to the twenty fourth floor

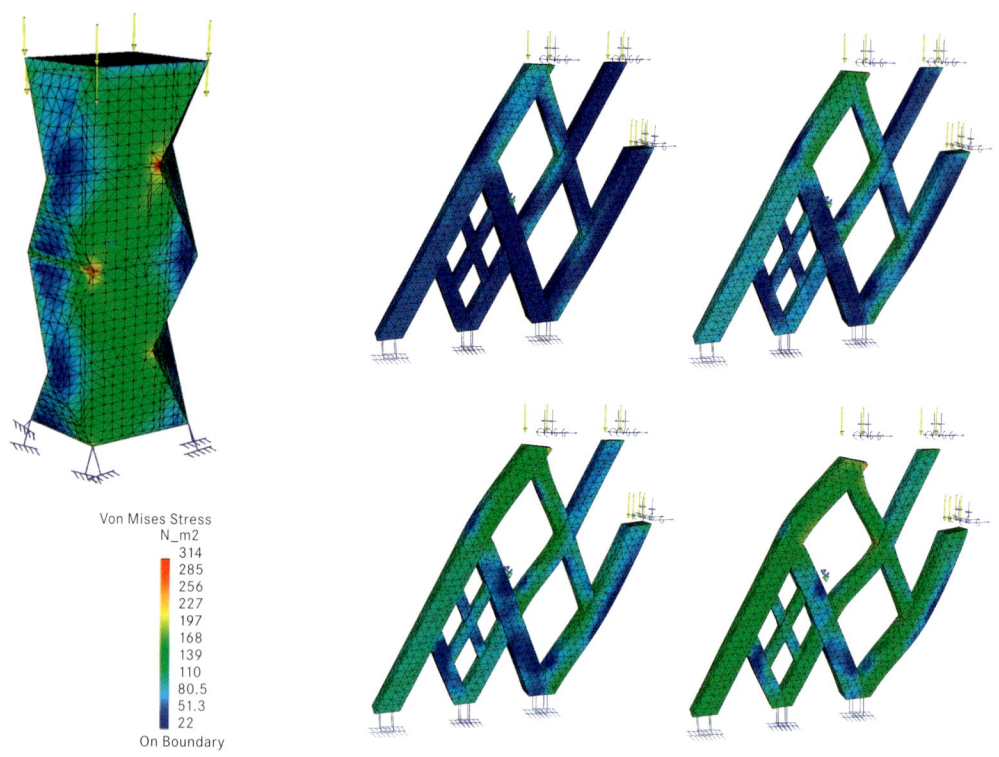

Von Mises Stress
N_m2
314
285
256
227
197
168
139
110
80.5
51.3
22
On Boundary

All connections and elements could be easily fabricated without resorting to uneconomic sections and nodes. FOA's setting up of a well-thought geometry, and the process of developing the structure and façade with us early on, assumed that a unique form that looks very different from every direction could be delivered.

Large commercial projects like this are well developed, so it becomes difficult to purposefully not do what had worked previously; historically, engineering innovations have tended to come from the public sector. With a slight change in process and a requestioning of what we know already, new forms such as this are produced. The contribution of the 'mega structure' on the outside of the building has produced an unexpected stiffness relieving the core from acting alone to provide lateral stability.

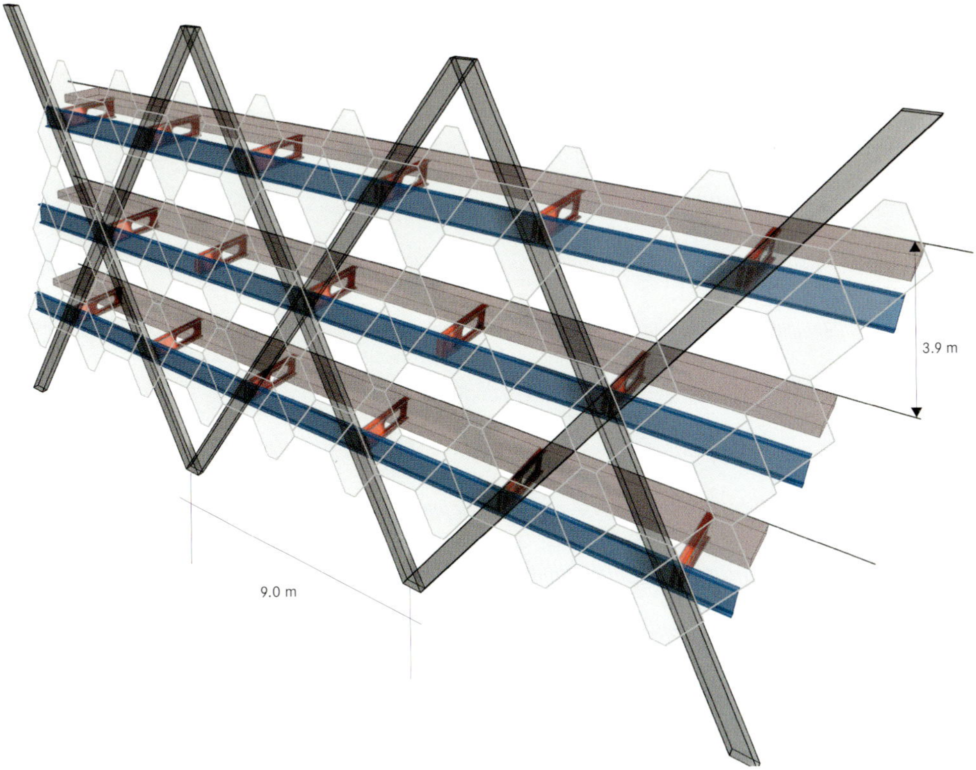

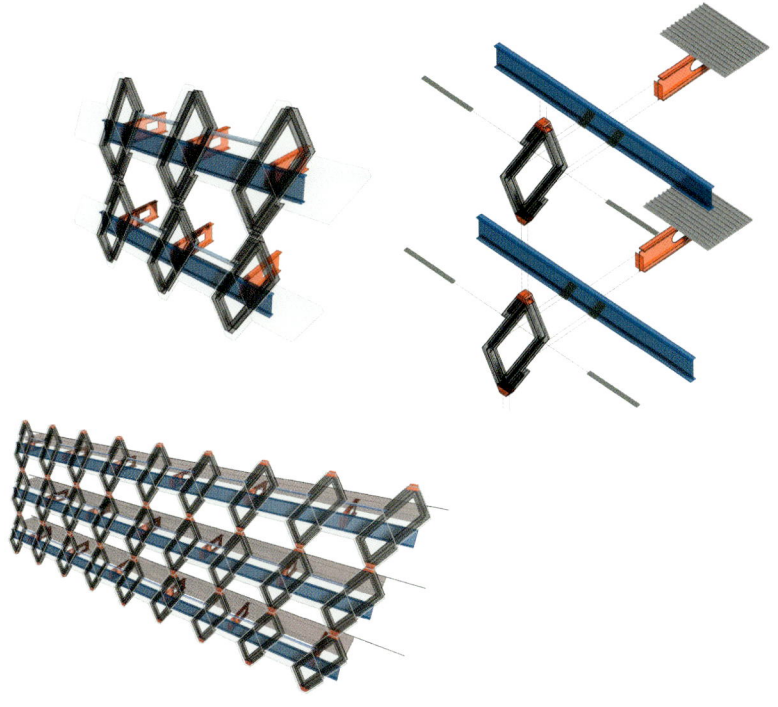

Façade: diagrid columns

Façade: stacked diamond columns

Nontraditionality

Alejandro Zaera-Polo interviewed by Hanif Kara

Alejandro Zaera-Polo is a founding partner of Foreign Office Architects and occupies currently the Berlage Chair in the Technical University of Delft

Hanif Kara is a director and co founder of AKT, currently Pierce Anderson Lecturer in Creative Engineering at GSD Harvard, Visiting Professor of Architectural Technology KTH Stockholm and a CABE Commissioner

Hanif Kara: Sometimes when we work with you we have to catch up quickly to understand what you're trying to reach. Is this because you strive for a particular vision?

Alejandro Zaera-Polo: Traditionally architects were visionaries. They depended on their own single vision of Utopia, which they saw as their duty to bring back to Earth and give it concrete form. But in the contemporary world this reference model of a single vision is no longer relevant. Now reality has become much more complex and in it we can find the interest for which we once had to construct Utopian visions.

HK: Working with you has made me realise that you tap into our knowledge of fabricators and it seems to inform your designs.

AZP: What makes Adams Kara Taylor interesting is the way they engage with this contemporary reality. Their pragmatic approach to knowledge sets the tone for their practice. They know what is going on in various branches of the industry and this intelligence informs the way they design. They understand, for instance, what steel-makers are doing and what they offer to engineering, while at the same time they appreciate how developers are operating in the market place. Bringing these different aspects of reality together establishes a datum point from which ideas can develop.

HK: I remember coming to your crits at the AA just before you won Yokohama. And then shortly afterwards we did Belgo's and that helped us as a practice shape a completely new stream of ideas. How we react is how you made us react.

AZP: Your have a quite different way of working from the visionary engineers of the past. You don't have big visions or big perspectives like perhaps Nervi had when he was working because then reality was much less mediated by technology. In figuring out how to make worlds of twisted steel or reinforced concrete they certainly had a grasp of reality. But everything they then would or could do was in steel or concrete. At the time the range of available materials and forms – and what you needed to know - were far more limited. Now there are so many parallel worlds colliding into reality that it no longer makes sense to be the "master of glass" or "master of concrete". It's much more interesting today to have direct knowledge of what steel fabricators or concrete manufacturers are able to do with their new machines than abstract concepts.

HK: I understand this point in the context of Nervi, because he had one fantastic idea and repeated. I can't do that. We now see so many more possibilities and demands, arising from the proliferation of knowledge.

AZP: Your skill is in "riding the wave", or surfing as Rem Koolhaas characterises the way we need to operate today. It is not our own vision which drives our practice but the proliferation of knowledge and the workings of the market. They combine to create a huge and unstoppable wave which we have to ride. Our skill as designers comes not from assembling ideas and forms that approximate to our vision of utopia, but in being able to make something interesting with what's there. Contemporary practice is about a permanently shifting and expanding range of options which derive from reality and experience rather than abstract concepts. None of the underlying conditions are stable, whether they are economic forces, the state of knowledge or relations between the different parts of the construction industry.

HK: When we started we set out to redefine the environment in which we could work. We realised we would have to have the right people in the office to take any particular knowledge at any particular time and employ it. Most previous engineers where only about problem-solving, all the time. We've tried to go beyond just solving technical problems, in your words to ride the wave.

AZP: On our first collaboration, the Belgo's restaurant in Ladbroke Grove you were able to apply their knowledge of the local construction industry to guide us through the project. We realised you are quite unlike conventional engineers in that you are neither abstract originators of concepts which address problems at a generic level; nor do you merely apply rules that somebody else has found in order to solve these problems. Is this really true that there were guys out there who were simply applying the rules that somebody else had found in order to solve problems?

HK: What's missing from them and differentiates us from them is the human dimension, a recognition that each project has particular contingencies which can only be solved by surfing the wave. Over the last ten or fifteen years architecture has gone through its own crisis. Engineers have started to look at themselves as well and we feel fortunate to belong to a generation that has learned to speak about the world, to say what you're doing and why you're doing it. That's something that traditional engineers did not need to do. Now we're asking as engineers how do we capitalise on these opportunities? How do we find people and projects where you can carry on questioning the discipline and questioning the way you process it. Each time the process is slightly different, which is a function of the condition we are all in.

AZP: This is not exclusive to engineers because architects used to be the same. Contemporary practice has a genuinely new level of complexity. In particular that complexity creates a paradox: the possibility of always being able to do something new translates into an obligation or at least an expectation that everything will be new, while there is also an increasing obligation to be accountable. In this situation engineers who really understand the implications of what is possible, who can navigate and test the various tributaries and corollaries of ever-expanding capabilities are very valuable. Just as contemporary architecture can no longer be held in thrall to a single utopian vision, so contemporary engineering can no longer rely on formulaic solutions. It has to seek out what is appropriate to the contingent circumstances of a particular project from the entire field of what is possible. For me now that exists in the territory that lies between professional liability and accountability and the demand for something special and unique that is now a common demand from the market place.

HK: Do you not think that shadows your creative edge? You have many layers of people to whom you have to explain what you've done. Does it not blunt your edge to be in the hands of people who don't necessarily understand?

AZP: Perhaps it blunts your edge but I am interested in the possibility that it doesn't. We are expected to do special buildings, but at the same time are accountable to an ever-growing body of people. I want to turn that accountability into something that is creative because it can drive things in a certain way, and because accountability and liability are the frames of reference for our work. As I do not want to be a visionary with a long-term perspective I want to practise in a rather different way to the vision of a corporate practice that emanates from Walter Gropius. This leads me to appreciate how Adams Kara Taylor work. I see you as craftsmen rather than 'visionary engineers'. You engage directly with a number of different processes and various flows that are pushing reality in a certain way. Within that you have the skill to give the wave a particular direction that optimises the ideas and the way you are given physical form. As an important engineer probably you are supposed to do special buildings?

HK: Yes

AZP: What these clients understand by special buildings is often something weird. But I am less and less interesting in daring projects of that kind, and more and more interested in order. For me, though, order means finding the optimum solution for the particular concatenation of contingent problems that a project throws up in such a way that the result, if not the process of getting there, seems obvious.

HK: Can accountability can be turned into a factor that is actually creative?

AZP: I can think of two of our collaborations, the entry for the Olympic velodrome and the tower in Busan, South Korea. In the second we a number of tendencies came together into a very beautiful and elegant solution. The client wanted a super-tall building and we were interested in a basic triangular form which was efficient on plan and gave appropriate orientation. At the base it started as a triangle, but progressing upwards transformed into a Y-shape plan. This change fitted the split between commercial space on the lower floors and residential above. So the diagrammatic requirements of the programme were resolved efficiently into the plan form of the building which could itself be made into a consistent structure.

HK: It was a very nice solution which you don't have to build as a velodrome. The process was enjoyable and something interesting came out of it.

AZP: The velodrome started with a picture I saw of a geodetic structure made from bicycle wheels – suggesting an obvious link between programme and structure. Taking that idea literally, though, became too expensive, so we and AKT developed a structure of tension cords which formed a diagrid around members which were in compression. The aim was to create an iconographic resonance between structure and function.

HK: When we work with you, you don't rely on a structural phenomenon or structural gymnastics to produce something special in each case. Sometimes you might, but you don't. The majority of architects would expect me, once I define myself as a design engineer, to come into the room and perform like a monkey and produce a gymnastic or special structure, something the architects can hang themselves to...

AZP: But combinations of ideas from engineering, architecture and economics can generate opportunities between themselves. For instance energy consumption might peak in different parts of the building at different times of the day, and perhaps these peaks can be offset against each other. So structural engineering comes together with mechanical and electrical engineering, while the architect brings knowledge of how issues like circulation and fire escape and be resolved within the typology of tall buildings. Each discipline uses the others in an opportunistic way to create a synergetic effect which amounts to a new prototype, and engineers can be in the driving seat just as much as architects. That sort of opportunism arising from the parameters of a given project interests me far more than being 'visionary'. This perhaps helps to indicate how it might be possible to redefine progressive practice within the construction industry without having to resort to a Utopian vision. That happens on many different levels, on technical levels, on political levels on cultural levels.

HK: It's much simpler when architects know what you're talking about.

AZP: I'm interested in being opportunistic, to find opportunities in a specific situation rather than to deploy yet again the same form or the same style.

HK: How do you explain Elizabeth House?

AZP: That's another case, a tower above London's Waterloo Station. We set out to embed the structural constraints throughout the concept for the project. Its form gave the tower different silhouettes, suggesting a pattern of shifting diamonds which became the geometry for the structure, an external diagonal lattice. It also set a pattern for the fenestration of hexagonal windows. This tessellated façade, clad in Portland Stone, generated a twisted tower which would balance a sense of solidity with a complex form.

HK: Even those who haven't seen this kind of image before recognise it as something special. I remember showing it to a guy from a large developer and he realised it was FOA. If he were to come to you for a tower he's thinking of, it would show how market conditions and market intelligence can work, because the market needs something that is not a replica...

AZP: All these developments exist within a context of the market. It can be very dangerous, but it is also about waves, and having knowledge of where things are going. That sort of knowledge can come through education, but education in an office as much as in an academy. The ideal is to create a culture in an office where intelligence is shared, where people put ideas on a table and others pick them up and discuss them. That discussion may need to be directed to make it efficient, but if it becomes a bureaucracy it will wither. Cultures and markets are opposed to bureaucracies. It is significant that we first met you at the Architectural Association where we were teaching, and which fosters this sort of free exchange of ideas. Your own studio, with many young engineers from a huge variety of cultural backgrounds, simulates this sort of environment.

COMPLEXITY

The increasing seamlessness of techniques and analysis across multiple parameters of a project can generate new problems and unforeseen complexities. With parametric techniques, all problems of a design become connected, such that unless the design is resolved at the smallest possible scale, nothing will work at the scale of the whole. In opening up new possibilities for integration and increased complexity in the relation between parts, the parametric is also increasingly undoing the separation between scales of resolution of a project.

The development of visual interfaces in computing allows engineers to control complex geometries that are no longer based on proportions and algebraic relations, but on approximation through calculus. NURBS, for example, are geometries determined by an infinite number of specific conditions, mediated by an inexact function rather than geometrical relation. In a world where the relationship between the generic and the specific is crucial, contemporary engineering technologies enable architects to design and manufacture organisations that are differentiated and yet consistent. Informal geometries, differentiated structures, and non-linear organisations are now controllable with the same degree of precision that traditional architecture controlled linear, proportional geometries. Mass customisation has come to replace the traditional opposition between one-off, artisanal manufacturing and mass production. Where historically architecture, landscape design and urban development were split between the picturesque and the rational, a third way is now possible that produces complex organisations with rigorous orders.

Engineering elegance

Patrik Schumacher

Partner at Zaha Hadid Architects, co-director of the Design Research Lab at the Architectural Association

Elegance has been promoted as a new watchword to guide the next step within the current cycle of architectural innovation.[1]

The elegance we mean is not the elegance of minimalism. Minimalist elegance thrives on simplicity. The elegance I am promoting here thrives on complexity. It relies on powerful ordering principles that can establish lawful and legible continuities within a given manifold. Elegance in our terms achieves a reduction of visual complexity, thereby preserving an underlying organizational complexity. In short: *Elegance articulates complexity*. This is my fundamental thesis, and I would like to argue here that a congenial structural engineering approach is absolutely central to this ambition.

Attributed to a person elegance suggests the effortless display of sophistication. We also talk about an elegant solution to a complex problem. In fact only if the problem is complex and difficult does the solution deserve the attribute "elegant". While simplistic solutions are pseudo-solutions, the elegant solution is marked by an economy of means by which it conquers complexity and resolves complications.

It is this kind of connotation that I would like to harness for a contemporary notion of elegance in architecture and engineering. An elegant building should entail an elegant structure and both together should be able to spatialize considerable organisational complexity without descending into visual disorder.

It is the sense of law-governed complexity that assimilates this work to the forms and spaces we perceive in natural systems, where all forms are the result of lawfully interacting forces. Just like natural systems, elegant compositions are so highly integrated that they cannot be easily decomposed into independent subsystems – a major point of difference in comparison with the modern design paradigm of clear separation of functional subsystems. In fact the exploitation of

---

1 See: Schumacher, Patrik, Arguing for Elegance, in: Castle, H., Rahim, A. & Jamelle, H., (eds), *Elegance, Architectural Design, January/February 2007, Vol.77, No.1, Wiley – Academy, London 2007*

natural forms like landscape formations or organic morphologies as a source domain for analogical transference into architecture makes a constructive contribution to the development of this new paradigm and language of architecture.

Structural engineering had its own significant share of inspiration from nature. Frei Otto went a step further and literally harnessed the lawfulness of physical systems as form-finding procedure to generate his design-morphology. The results have been striking. These processes might be referred to as "material computing"[2]. Such analog form-finding processes can complement the new digital design tools that might in fact be described as quasi-physical form-finding processes.

Elegant compositions or complexes are highly integrated formal/spatial systems that look like those highly integrated natural systems where all forms are the result of the lawful interaction of physical forces or like organic systems where the forms result from a similar play of forces selected and integrated in adaptation to performance requirements. Such elegant compositions resist decomposition, just like their natural models.

COMPUTING ELEGANCE – THE DIGITAL REVOLUTION IN ARCHITECTURE AND ENGINEERING

Current digital modeling tools are able to facilitate integrative effects: lofting, spline-networks, soft-bodies, force-fields etc. Morphing – the ultimate effect of animation movie technology - has been an often emulated paradigm for achieving the continuity of the differentiated.

There is an inevitable, powerful relationship between the new digital tools on the one hand and the new organizational patterns, compositional tropes, stylistic characteristics and aesthetic values on

---

2 *What Frei Otto called "Formfinding", Lars Spuybroek refers to as "Material Computing" in order to emphasise the similarity of those physical processes with the by now familiar and ubiquitous digital modelling techniques offered by animation software like Maya.*

the other hand. At Zaha Hadid Architects we are currently promoting the slogan: *Total Fluidity across all Scales*.[3] In fact it has become increasingly easy to achieve abstract sketch-designs (surfaces) that satisfy this slogan and thereby achieve a measure of elegance as defined here. However, pure geometry (surface) models are only the first sketchy step in the design of an elegant architecture. Only in limit cases such as the installation "ice-storm"[4] – an extensive experiment in morphing - does the modeled surface translate directly into a built reality. (The whole environment was executed as a CNC-milled poly-styrene body with a hardened poly-urethane skin.)

It is quite a different challenge to maintain a high degree of fluidity and coherence with respect to the design of a fully functional building where multiple functional and technical requirements impose the handling of multiple material sub-systems: the (multi-layered) envelope system, the system of internal spatial divisions, the system of circulation/navigation, the various service systems, and the structural system. The necessity of distinct subsystems poses the crucial task of organic inter-articulation as the battle-field where elegance is won or lost. The structural system often plays a domineering role in this concert of subsystem. *Organic inter-articulation* might be achieved by mutual geometric affiliation, inflection or by means of establishing lawful correlations between the various patterns of differentiation that are specific to each subsystem. The contemporary desire for smooth transitions and gradient transformations between conditions might be correlated with gradient structural modulations that translate the continuously differentiated distribution of forces that operate in any structural system.

STRUCTURAL FLUIDITY

The digital revolution that brought a series of powerful new design tools into architecture has also provided structural engineering with new tools to analyse and calculate structures in the manner that is congenial to the architectural ambitions towards total fluidity that have been unleashed by the new design tools. Traditional architecture was a game of assembling simple platonic forms like cubes, planes,

---

3 Schumacher, Patrik, Digital Hadid - Landscapes in Motion, Publisher: Birkhaeuser 2004. Zaha Hadid & Patrik Schumacher, MAK (Museum for Applied Arts), Vienna 2003

4 The Phæno Science Centre in Wolfsburg - completed in 2006 - was designed by Zaha Hadid Architects and engineered by AKT.

grids, domes, and pyramids. The key characteristic of contemporary architecture that challenges engineering is the pursuit of complex three-dimensional geometry and continuously changing forms. Such forms can no longer be analysed by means of decomposing them into discreet systems. This is significant because it challenges structural engineering with respect to its most basic concepts.

Traditional structural engineering relies on the ability to decompose any structure into clear and independent structural subsystems. Each sub-system adheres to standard concepts like column, beam, portal frame, arch, slab, vault, framework etc. Each of these concepts is characterised by a clearly typified geometric schema with its attendant distribution of forces. Within each simple subsystem the active forces can be easily ascertained, and great care is taken to control the transference of forces from subsystem to subsystem by the precise articulation of the joints. The overall arrangement of forces can then be traced step by step. This strategy of clear and distinct decomposition sacrifices efficiency and redundancy for analytical clarity and tractability. It is a strategy for the reduction of complexity that recognises the narrow computational capacity of the pre-digital era. In contrast contemporary architecture creates spaces which are morphing different spatial sections into a seamlessly differentiated continuum that resists such decomposition. In all these traditional systems the ability to analyse and calculate the behaviour of the structure is premised upon the purity of structural type and the severing of all redundant connections.

It is precisely the underlying typology – the thinking in clearly defined types – that is disappearing from contemporary architecture. In fact, "From Typology to Topology" is one of the key slogans of contemporary architecture. This implies that contemporary architecture escapes all traditional engineering procedures. Within a contemporary avant-garde building like the Phæno Science Museum in Wolfsburg[5], the structural systems morph as much as the architectural forms.

With new engineering tools like finite element analysis, which break the structures into particles rather than into parts, the engineer

5 The Phæno Science Centre in Wolfsburg - completed in 2006 - was designed by Zaha Hadid Architects and engineered by AKT.

is able to capture the ever shifting arrangement of forces. The universe of potential force patterns becomes boundless.

"From Parts to Particles" is another key slogan of contemporary architecture.

Structural engineers can now analyse mixed, hybrid systems. A tool like Finite Element Analysis can also cope with dense, redundant interrelations of the parts of a structure. We no longer need to sever and isolate the structural components or subsystems. This means that we can harness the structural efficiency of an interconnected network, where parts work together rather than remaining independent from each other. The re-tooled engineer allows the structural forces to flow freely through the surfaces provided by the architect. This is the era of structural fluidity.

In the case of the Phæno Science Museum (to be described below) we can observe a mixture of spanning, cantilevering and vaulting within a waffle slab whereby spans and cantilevering dimensions are continuously changing. The cones flare into waffle-slab rather than remaining discreet props that pick up their load at distinct points of contact. The space frame above is continuously differentiated whereby each member within the space frame has a different angle (the grid fans in two directions) so that each cell of the space frame has a different size. In a complimentary move each member has a different thickness and weight. Obviously, this nuanced optimisation can only be coped with by means of computers, both with respect to the calculation of forces as well as with respect to the handling of the geometry and manufacturing schedules.

EMERGENT PERFECTION
The notion of elegance promoted here still gives a certain relevance to Alberti's criterion of beauty: you can neither add, nor subtract without destroying the harmony achieved. Except in the case of contemporary elegance the overall composition lacks this sense

of perfect closure that is implied in Alberti's conception. Alberti focused on key ordering principles, like symmetry and proportion. These principles were seen as integrating the various parts into a whole by means of setting those parts into definite relations of relative position and proportion in analogy to the human figure. Perhaps the best example of this ideal is the Palladian villa. In contrast contemporary projects remain incomplete compositions, more akin to the Deleuzian notion of assemblage than to the classical conception of the organism. Our current idea of organic integration does not rely on fixed ideal types. Neither does it presuppose any proportional system, nor does it privilege symmetry. Instead the parts or subsystems mutually inflect and adapt to each other achieving integration by various modes of spatial interlocking, soft transitions at the boundaries between parts, morphological affiliation, and lawful correlation between parallel patterns of differentiation etc.

Naturally, on the way to the elaboration of fully functional, fully detailed designs, whereby evermore systems or layers need to be integrated, the principle of inflection (organic inter-articulation) becomes evermore difficult to maintain. Also the visual field is in danger of being overcrowded, compromising legibility and orientation.

It is at this moment of mounting difficulty – in the face of bringing the new paradigm into large scale realization – that elegance becomes an explicit priority, not least because the built results have all too often been disappointing in this respect.

The principle of elegance postulates: *do not add or subtract without elaborate inflections, mediations or interarticulations*.

While the classical concept of preordained perfection has thus been abandoned, there still remains a strong sense of increasing tightness and stringency, approaching even a sense of internal necessity, as the network of compositional relations is elaborated and tightened. This network of "compositional" relations includes the arrangement and morphology of the structure.[6] I guess every designer knows this from his/her own design experience. The more the compositional cross-referencing, inflection and organic inter-articulation within the design has been advanced, the harder it becomes to add or subtract elements. This kind of design trajectory – although wide open at the beginning – beyond a certain point becomes heavily self-constraining. One might be inclined to talk about the increasing *self-determination* of a composition: an emergent (rather than preordained) perfection.

I guess every designer knows how a design-trajectory can lead into a dead end, can fail to "work", or remain unresolved. The elegance we mean – elegance on top of complexity – is a tall order, and can not be secured in advance.

With increasing complexity the maintenance of elegance becomes increasingly demanding.

Complexity and elegance stand in a relation of precarious mutual amplification: a relation of increasingly improbable mutual enhancement, i.e. mutual amplification with increasing probability of failure.

The recently completed Science Museum in Wolfsburg ("Phæno") is the virtuoso masterpiece in the articulation of complex continuities that can be followed all the way through the building. The whole building is inscribed within a rigid trapezoid whose angles are adapted to the site-condition. Within this sharp-edged trapezoid everything flows and melds without corners. The ground-surface is molded into an artificial topography that registers and receives the cones that carry the building. These cones – executed in insitu reinforced concrete - constitute the primary structure of the building. Each cone has its own variation of angles and radii. The cones blend seamlessly into the waffle-slab above. Some of those cones also reemerge within the interior – either as craters or as cones that continue to carry the space-frame above. There is an essential symbiosis in the spatial and structural conception of the building, and a close inter-articulation of the waffle concrete structure of the raised floor and the steel space-frame that carries the roof. To a large extent the architectural expression is dominated by the structure. In fact, the structure constitutes the architecture, and therefore the demand for tight collaboration was extraordinary.

---

6 *In fact we have to admit that for us architects the structure enters our considerations as just another set of compositional elements.*

The lateral openings are of two kinds: the large openings are conic sections that produce the characteristic parabula form, and the smaller openings come in swarms that are articulated as variations of the swarm of voids that make up the waffle slab. In both cases the openings closely relate to the respective structural logic of the surfaces penetrated. These openings in turn contribute and relate to the rhythmic flow of the interior spaces. The large, continuous expanse of space on the interior is captured between the crater-scape produced by the cones pushing through the floorslab and the continuous space-frame. These two layers are correlated via mutually echoing shifts in section. It is this resonance between the various layers and sub-systems that gives this space its exhilarating sense of complex order that we perceive as elegance. The fact that this stimulating spatial experience is delivered by the structure itself rather that by some less substantial and more lightweight layer adds enormously to the power of the effect.

Discourse networks and the digital
Structural collaboration at the Phæno Science Centre

Tim Anstey (reproducing a conversation with Hanif Kara)

Tim Anstey is an architect, researcher and lecturer at the University of Bath Department of Architecture and Civil Engineering.

The Phæno Science Centre in Wolfsburg, like many "centre stage" buildings, is inevitably assimilated into architectural discourse in terms of the authorial contribution of a famous architect. Yet the ways in which such complex buildings are understood, and the extent to which that authorial ownership is annexed by or is spread out from the single "name" architect they are attributed to, varies widely. In that distribution Phæno is important in several ways. Although the structural thinking of the project is one layer within a complex whole, it is clearly a highly significant one and at the heart of the architect's years of experimenting. The building cannot really be explained, and certainly could not have been conceived, without acknowledging the way in which structure and structural engineering is used as a means to develop architectural ideas, and discussions about the building have tended to stress the collaborative nature of the project. It is evident that the habits of architecture/engineering collaboration that exist in London, which is where the thinking that created Phæno was done and which it exploits, are unique. The discussion that began during the 1950s following the establishment of Arup and Partners (1946) and F.J. Samuely and Partners (1956), created a discourse network that fundamentally effected how architecture was thought in Post War Britain. In this the context of the technical – an invisible topography of contractual, physical and production conditions surrounding architecture as object – came to inform architectural action in new ways. Yet even in this culture, Adams Kara Taylor collaboration with Zaha Hadid Architects on the Phæno building offers an unparalleled integration.

Hanif Kara believes that one of the things that has both been promoted by this culture of engineering/architectural collaboration and has served to develop it is the active role of the more experimental engineering offices in architectural education. The discussions that produced Phæno relied in fact on its architects' and its engineers' shared experience as studio tutors at the Architectural Association School of Architecture. At one level this type of academic role invites the pursuance of ideas for their own sake, and appears particularly important in creating a platform on which architectural and engineering culture can meet. At another it alters the boundaries for "problem solving" between architects and engineers and sets up a very fluid condition for the interchange of information. At Phæno this fluidity allowed the thinking systems that condition the building to be reclassified. The ten reinforced concrete cones that are central to the building's integrity are not quite one thing or another; they are part arch and part beam, but in order to analyse them both structurally and architecturally, one has to get away from the traditional ideas of thinking and naming. The result is more about topology than typology.

Yet that academic context – a context of talk, air, possibility, not limited immediately or necessarily ever by the current envelope of technical possibility – has to meet another kind of discipline in order to change actual building construction and procurement, and to produce buildings like Phæno. Kara's maxim – we shouldn't teach technology, we should change it – emerges from the openness of discussions in academia and from the hard experience of practice. One of the most impressive aspects of AKT practice is their willingness to take on the grinding technical research that's necessary to support architectural ideas without grounding, wounding or killing them. This research must be carried out at the level of the material analysis (what material will do the job, and what do you have to do to an existing material to make it do the job?) and at the level of production and economy – research that will vary from project to project.

The organic process of form making that defines the hybrid elements of the building – concrete and steel – produces questions about buildability that require a separate formal and creative analysis exercise. While the building is conceived to be stable as a complete assembly, its parts – particularly the reinforced concrete cones that organise the building – are unstable as separated elements; so a procedure of analysis had to be undertaken to vouchsafe the "history" of the construction – to ensure that the building was stabilised at each point in its physical development. This need both to define avenues of investigation and to pursue them is also evident at the material level, thus there is a whole narrative around finding a self-compacting concrete that could be cast into the complex forms developed for the lower parts of the building and to test as necessary to ensure the general product will work in the local conditions of the design. The whole process involves a delicate balancing act in which ideas are allowed to remain sovereign while research into problem solving is pursued as far as possible.

Underwriting all these various forms of research is the basic level of digital analysis and form making – the production of digital tools that have a capability to analyse problems in new ways, as a means for interfacing with a discussion of architectural possibility. The way in which digital information is now exchanged between consultants challenges the traditional nature of drawing and representation that emerges around architecture. If the drawing has, through history, acted as a prosthetic device partly to guarantee the authority and genius of the architect – if architectural drawings have always had difficulty in escaping that rhetorical dimension – the nature of digital information exchange can be seen as a significantly freeing gesture.

In terms of disciplinary collaboration, Kara explains that the impact of digital technologies must be seen to be fundamental, partly in that it challenges existing ways of representing and explaining projects, partly in that it creates another kind of "space" in which the thinking about projects might be done. At the Wolfsburg Science Centre this exchange relied on finite element analysis software developed by SOFiSTiK, a German software company specifically for the project. Again, the research story is important. At the outset of the project no software existed that could precisely analyse the combination of form and material that was emerging from the design, although SOFiSTiK's almost could; thus AKT performed a tailoring role by setting the parameters for task-directed development of the software company's existing product. Having evolved the product, it then became central both for the structural and architectural design. SOFiSTiK could model continuing iterations of the form of the concrete cones and provide both analytical information about stresses and a model of the formal result back to the design team; material was slowly edited out of the design through this process of subsequent testing, a process that appears to have empowered both the architects and the engineers involved.

The creation of the "thinking space" that emerges through using digital tools clearly fascinates AKT. It is also clear that this need to create a shared space has to impact on the nature of the digital project. The significance of the digital is that it provides a space that architects, computer scientists, specialised analysts, material technologists and engineers can share in order to translate and enhance the first ideas in a project. The aim of this is not necessarily always to "optimise" a design, but to find a process that departs significantly from the ones both architects and engineers are trained in. The environment then demands "new tribes" to be involved in this in-between space, which is also about research on projects – no one pays for it, so it has to be part of the work. The more complex the tool, the less fruitful the collaboration. This is because a very limited number of people can use specialist complex software; so it needs special projects and specialist operators. Writing project-specific software is a trend but is problematic in terms of intellectual property rights. Where possible, AKT are trying to enhance and use tools that are easily available. Another issue is the gap between design, manufacture and site, as not many constructors are able to deal with the latest software and tools.

The experience from Wolfsburg, then, must be essential in explaining the dedication with which AKT have pursued the goal of producing design software that combines architectural modelling, structural analysis and evolutionary characteristics. And if the collaborative "space" provided by digital technologies permits the integration of geometrical and structural analysis in new ways, it is clear that this is only one aspect of its significance. At the same time it changes traditional roles in relation to project representation, and allows other areas of analytical expertise to enter into and impact on the arena of the design. These possibilities exist both at the level of design analysis – predicting how materials will behave in complex situations – and at the level of production. For example, digitally driven processes, such as printing concrete, could potentially revolutionise both formal possibilities and building production methods.

That desire lead AKT to set up p.art in 2000, an in house research arm which looks to the development of structurally intelligent formal modelling software. It is about geometry as just one strand; yet there are others that are critical. The "engineering software" is what the company seems to be engrossed in. Trying to make it perform quicker and better becomes a lay task in this area. Kara observes that in the company's case, the p.art vehicle also deals with off-stream work like artist collaborations or specific material research. They also seem to be "judged" on the way their work is presented, which is a situation familiar for architects but not for engineers, who tend to distrust rhetoric as a non-scientific area outside the scope of their work. It is important to stress that the scale issue is such that there is a limited pallet of materials for building structures. Timber, steel, stone, concrete and glass are likely to remain at the centre of architectural production. However, there is now the possibility for the company to take a "forensic" look at the nature of these materials and to use them in radically new ways. Both the concrete and the roof structure of Phæno make use of this analytical potential to change material use. Here a "bespoke" use of the two common materials is facilitated by latest CAD/CAM methodologies. This kind of investigation demands the cross-disciplinary expertise and flexible working structures that p.art provides for.

What are the tensions that such initiatives set up? Are there issues to do with secrecy versus openness when part of the work of the office becomes classified as research? Does the creation of new forms for consultancy around structures also speak of new possibilities for project ownership? An intriguing subject for study, in this context, is the issue of how we are to describe the significance of interdisciplinary and engineering culture for architecture.

In Kara's mind, the context of what architects call "research" is very different from engineering research, and that the two disciplines should keep their separate identities as the basic thinking is different:

one has the ability to diverge whist the other wants to converge. But he believes that both should be educated in the other's discipline – so they know what to ask and when, and that research should be shared. His ideal ambition is that industry can begin to think outside the box in relation to the secrecy that surrounds new research. This does seem possible with developments around digital technology that provide a common platform.

And what can't the digital do? Clearly, digital technology and in particular, analysis is here to stay, but it has a dark side. The homogenous nature of some digitally-driven projects, poured out of Maya, Rhino, etc., creates a shared illusion of reality that then risks separating it from creativity. The company should not simply use the latest technology for its own sake; to do so would be to produce "monsters". Thus in self-interrogation, as well as interrogating project briefs, clients and other consultants, it is important to ask "Why?". Also, human transaction is the most crucial part of any creative process. For example, after all that hard work on Phæno, the basic component of the concrete is the rebar (reinforcement bar), where it is placed and how it is fixed in place. No amount of digital analysis will remove the basic requirement for defining, placing and monitoring the processes by which large structures are erected.

Perhaps, then, most of all this fruitfulness comes out of an acknowledgement of the social and its significance. What Adams Kara Taylor perceives as special about themselves, is not that they are experts in Digital Projects but that they are multicultural, both in terms of discipline and ethnicity. "We speak 31 languages here; we make a pluralistic difference". And this self-image is perhaps related to a broader instinct about the importance of the human. Leon Battista Alberti praised Filippo Brunellsechi for the social achievement of vaulting Florence's Duomo, suspending ponderous masses "on air" and creating a structure large enough to cover all the people of Tuscany in its shadow. The social organisation required to achieve such feats will always be just that – social.

The concept behind Zaha Hadid's Phæno Science Centre is to simultaneously de-mystify science and unite Wolfsburg's severed urban fabric. Ten cones rise to support the elevated form, morphing gradually so one surface warps and curves to form the entire volume. The design breaks standard conventions of wall, roof and floor, and required a new approach to structural analysis. Collaborating very intensely with Hadid's office, AKT tested the outer limits of new analytical software to understand the building as a complete form, rather than a series of discrete elements each with its own forces.

The design leaves no scope for redundancy. All the elements work together; the walls and slab combine to make a continuous shell. Using the SOFiSTiK finite element software, AKT tested every part, including each of the varying steel joints. Without this tool, the structure would have been more conservative. Another innovation involved the use of self-compacting concrete, where a chemical additive ensures that the concrete retains the correct mix without the need for vibration, allowing AKT to more accurately predict its overall performance.

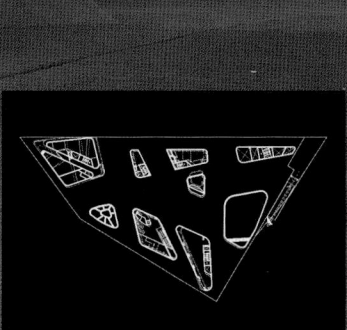

CONES
PHÆNO
SCIENCE
CENTRE

ZAHA HADID ARCHITECTS
WOLFSBURG, GERMANY, 2005
CLIENT: CITY OF WOLFSBURG

Finished Roof Structure

This is a project that is actually about seamlessness, which seems to create a new temporal problem for a building: the issue of a system designed to be stable as a whole, but not when its elements are considered individually.

Literally, the project only stands up when the ten concrete cones are finished; if one cone is left out, the whole thing collapses, whereas in a normal building, the removal of a single column will not lead to catastrophic failure because there is a certain redundancy within the column grid. At the Phæno Centre, because of the seamless nature the different parts of the project – programme, space, analysis, or more properly, engineering and architecture – those kind of traditional redundancies could not be realised.

The complex geometry seems to have been the main source of innovation. How did the cones originate?

The scale is perhaps the best place to start from as, at 154m × 130m × 97m, the overall size demands that the structure has sufficient vertical support. Through considering the urban strategy, the architect located what later became the 10 cones. These structural components were developed as inverted truncated-cones. Each has a geometry consisting of the wall areas connecting by curved geometry at the corners. The walls of the cones continue down from the ground plane to the basement with inclines ranging from 35° to 90°. The cones rise up to support the main floor and only four continue through to the roof. This created 'complexity' of various orders ranging from how one draws the cones, to how one reinforces and constructs this on-site. This level of complexity can only really be dealt with by the use of the most advanced digital technologies. The temporary condition as one builds these bottom up was also challenging as the temporary supports and stiffening systems (Doka Framax and Dokaflex) had to be invented by the concrete contractors Heitkamp.

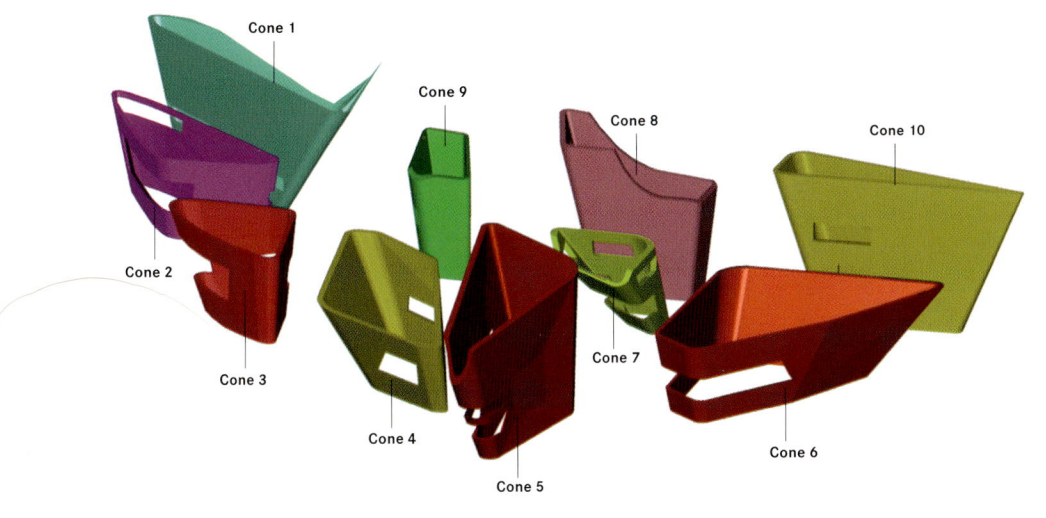

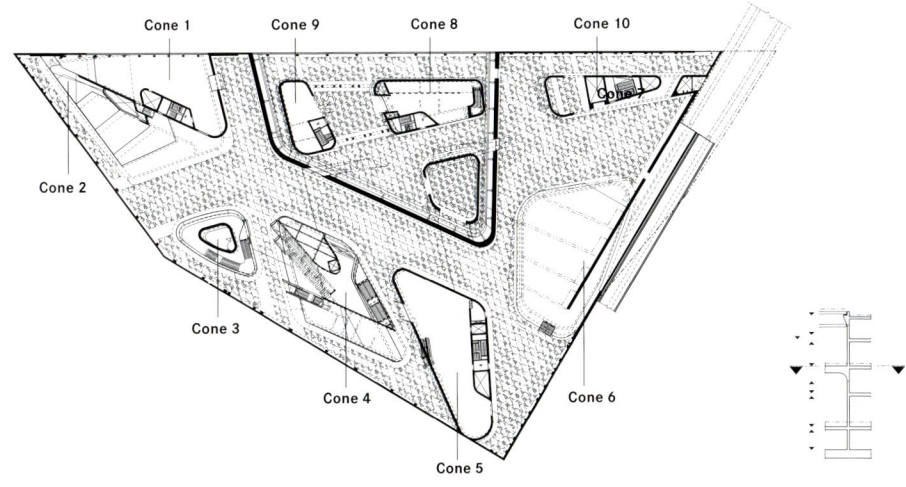

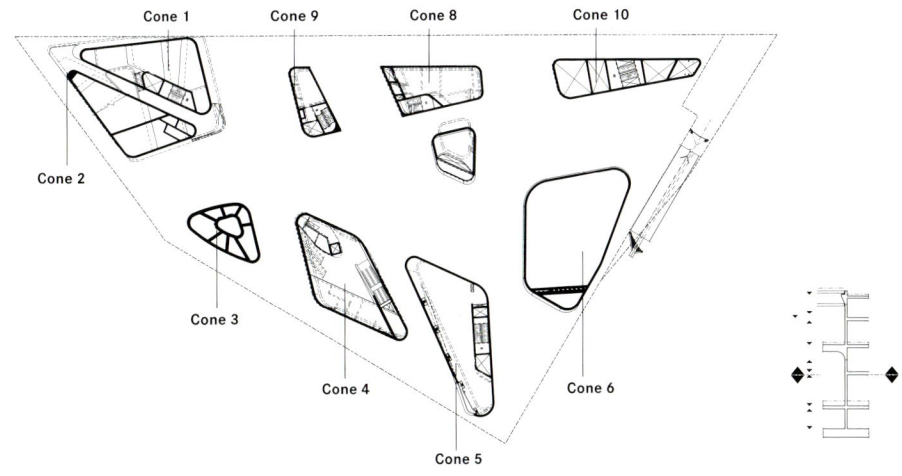

Arrangement of Cones

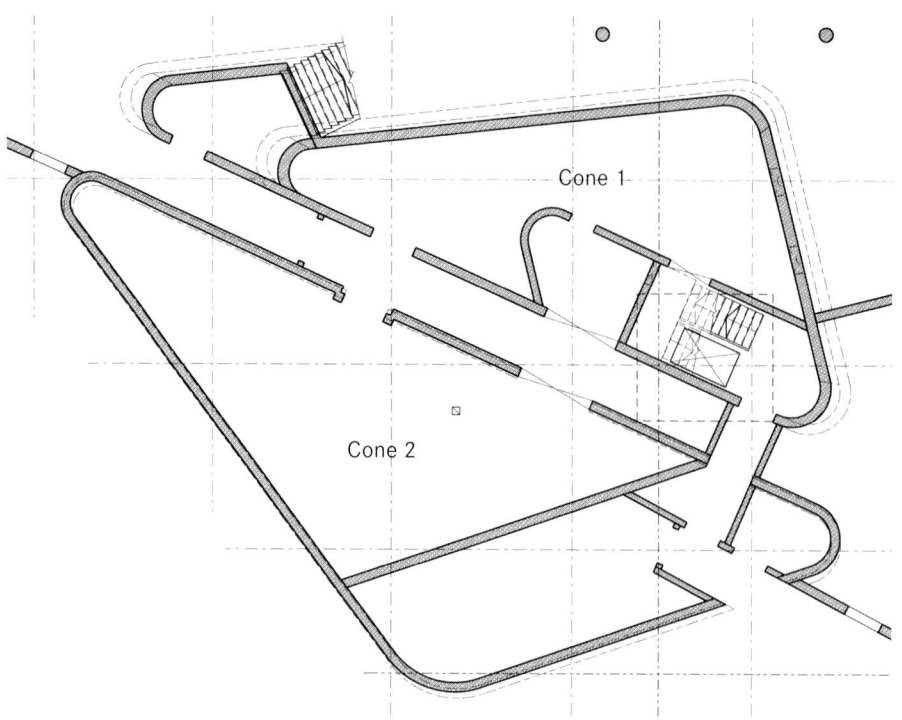

Setting out tangent points of cone 1 and 2 in plan view at basement level

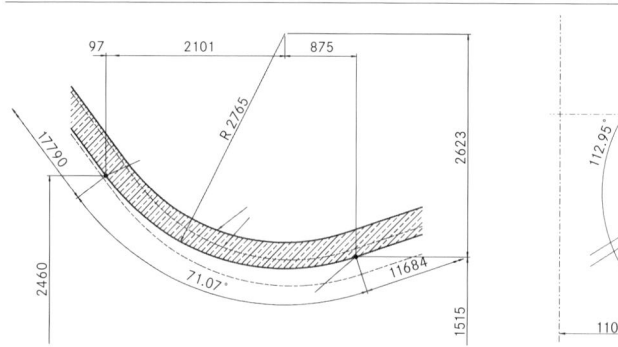
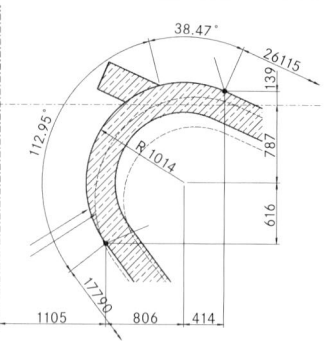

Corner detail of setting out tangent points cone 2

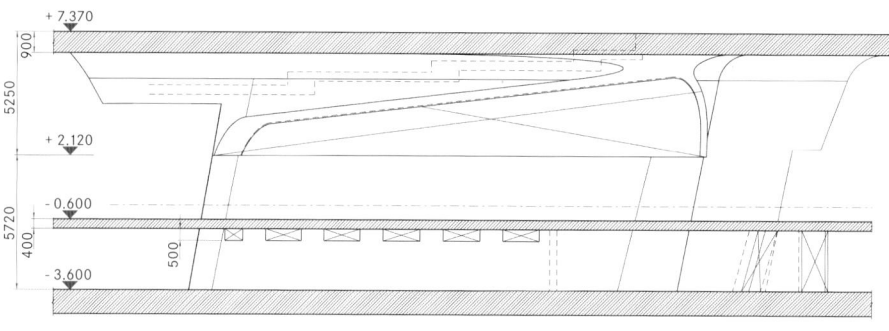

2-dimensional section drawing cone 2

Temporary works during cone construction

Cones prior to completion

Cones completed

Going back then, how were the cones finished?

The architects carefully specified a variety of finishing with planks to be cut conically and anchorage points in the centre of planks only. But the most interesting junction is where the cone meets the slabs - a 'seamless' solution was used by the contractor using GRP elements, as this junction is genesalli of small radii and doubly curved. The formwork thus was a project in itself since the finishes and shapes were carefully prescribed to achieve the best exposed concrete.

The Southwark Station project is a 520-square me-tre conical wall composed of 496 triangular shaped panes of blue glass. It is the outcome of collaboration between architect Richard MacCormac, artist Alexan-der Beleschenko and AKT. Enclosing a subterranean level between upper and lower sets of escalators, the geodesic surface is a simulacrum of the sky.

To determine the glass array, AKT devised a software programme which "unfolded" the cone into a series of triangles and which fed straight into the fabricator's cutting machine, keeping tolerances to the minimum. The pieces are attached to a simple steel frame by a custom stainless-steel, six-fingered connector.

AKT designed the product to hold each piece of glass in its correct position, and also developed a software programme that aided fabrication. Each finger is standard, but is free to slide and turn to adapt to the different conditions.

TRIANGULAR GLASS+SPIDER CONNECTIONS SOUTHWARK JUBILEE STATION

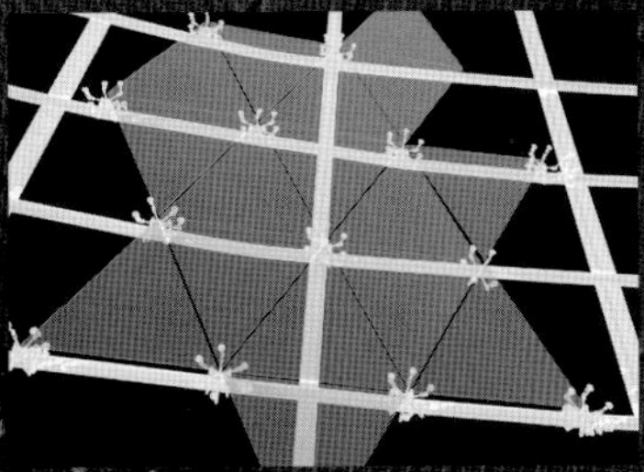

MACCORMAC JAMIESON PRICHARD
ALEX BELESCHENKO, artist
SOUTHWARK, LONDON, 1999
CLIENT: JUBILEE LINE

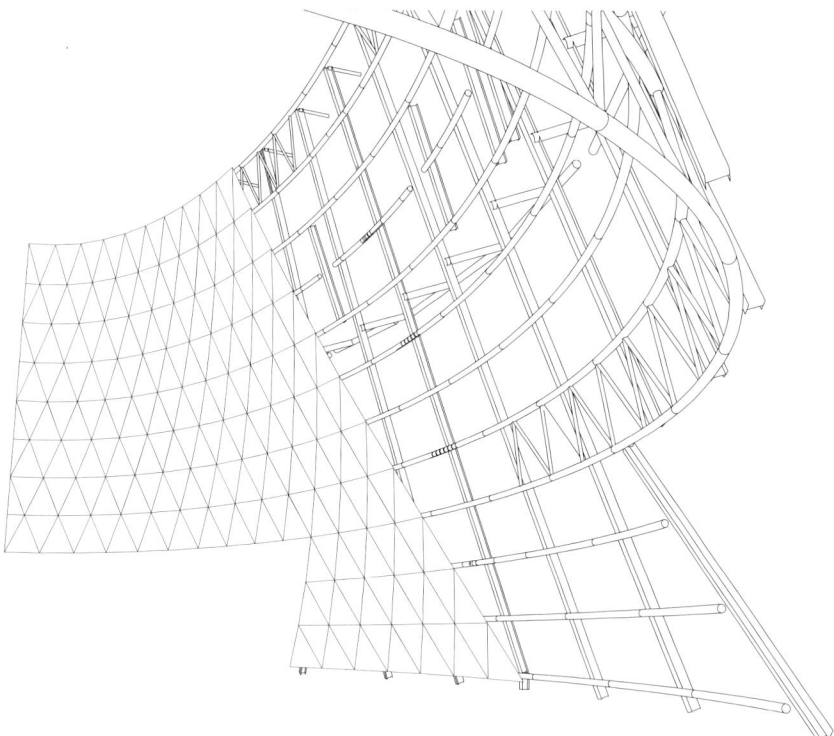

Diagram of wall with structure and glass

The project exemplifies the role of art in both architecture and technology. We began this project at a time when geometric modelling became a central curiosity for our work and changed the way in which our discipline can practice. It can be seen as a predecessor of the work that our p.art group is doing today. The project came at a moment when the term 'parametric' design was being ushered into the new vocabulary of architecture. Designing a leaning B-spline curves was a new thing, and we had to figure out how to draw and automate the making process.

We were trying to create a geometric model to develop the following problem: how to support a node that is comprised of six triangles, whereby each triangle is unique due to tiles changing along the spline curve? The geometry was certainly achievable, as the unfolded triangle drawings were all created digitally. However, the key point was to enable the structure to work, as well as fixing the triangle into position.

In our solution, the steel structure behind the triangulated glass is very simple. What is innovative is the transition between the triangle and the steel. This is achieved through the use of a six-arm "spider" bracket that picks up the geometry at any point, except when we had to cut in the corners. Here, the spiders have only four arms. The design uses distinct parameters such as the need of holes in a triangle, whether the triangle changes shape as it curves, and the thickness of the glass.

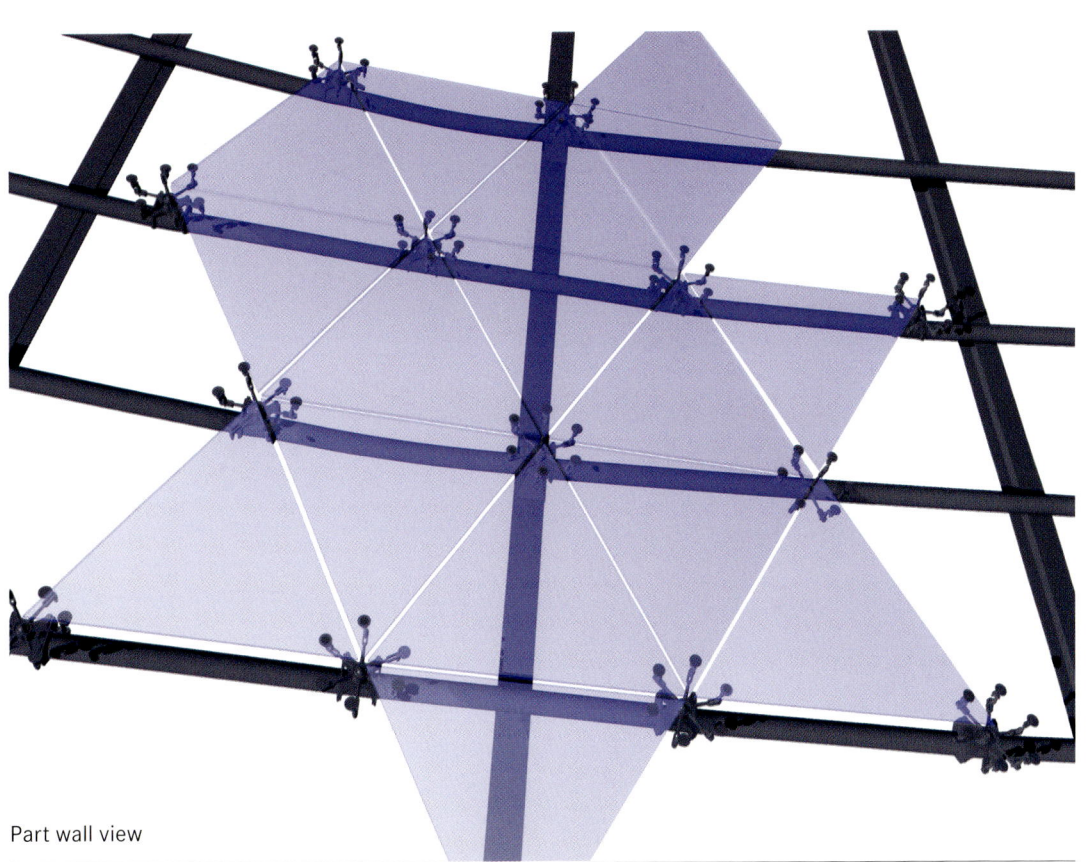

Part wall view

So the thickness of the glass actually changes from panel to panel?

It changes in three tiers due to size, and also because of loading derived from use. The first four metres of the cone are designed for crowds because it's a railway station, but the thickness is then reduced on top to make it cheaper.

What are the factors in the loading? Is there also wind and air pressure?

There is suction, which changes from the bottom to the top. So we had to deal with the analytical factors forensically, and simultaneously think about the "knuckles" at the junction in the glass (as the weakness in glass systems is always the junction). Where there is a hole in the glass, there is a knuckle in it, which is a rotating device that determines the weakness or the strength of the whole system, both as a structural fixing and as a device to allow rotation.

The supporting device is made up of an L-shaped arm and the knuckle. One part of the arm moves up and down the slot, while another part lies on the other side of the knuckle and rotates. Therefore, it picks up a whole envelope—a circle and another circle of geometry. Thus the hole in the glass can be anywhere in this space, in the whole geometry, and this device will support it.

Unfolded geometric mapping

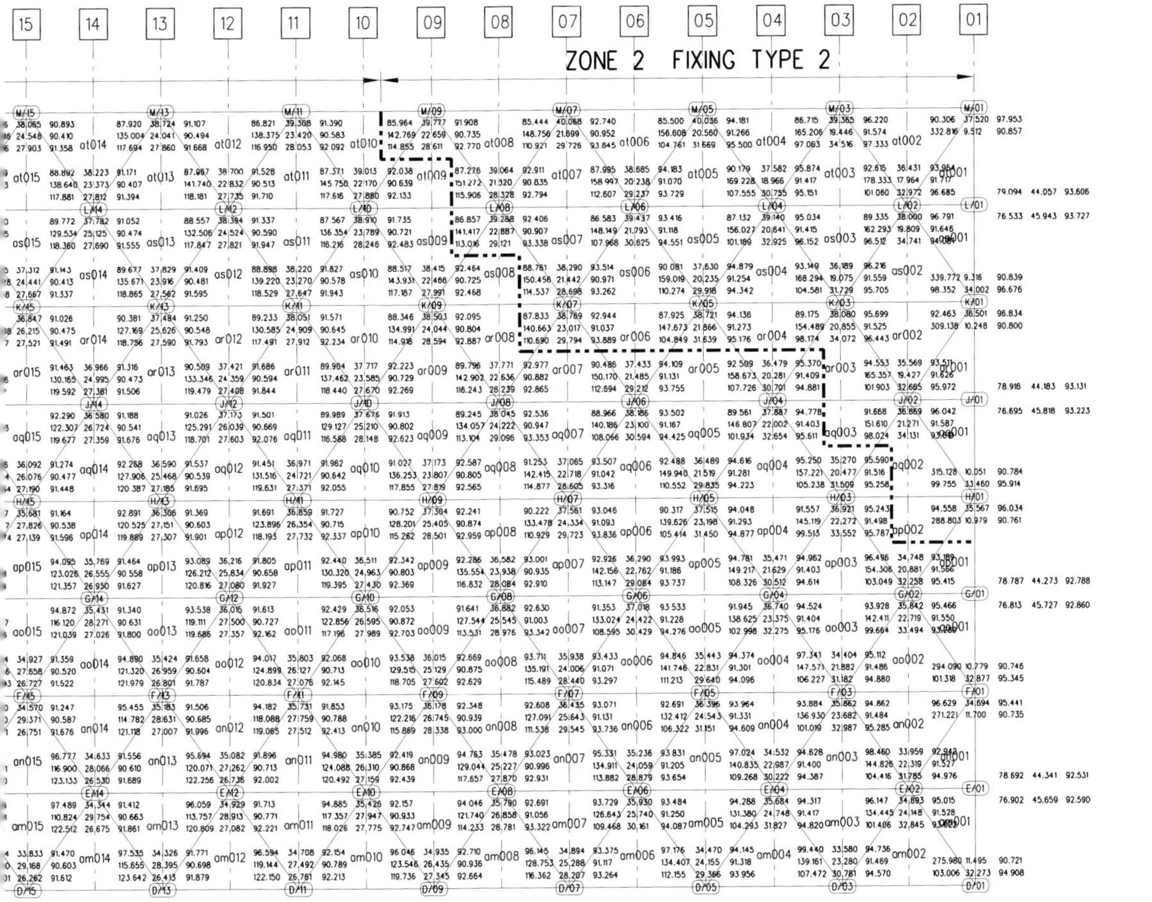

ZONE 2   FIXING TYPE 2

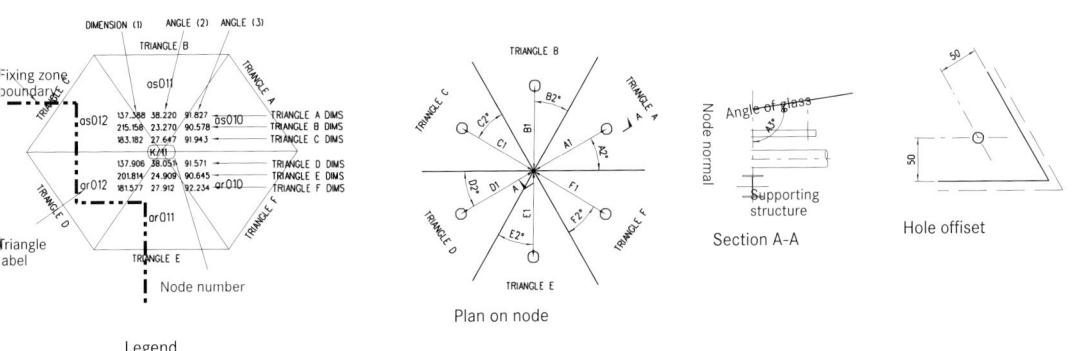

Legend

Plan on node

Section A-A

Node normal
Angle of glass
Supporting structure

Hole offset

Fixing zone boundary

Triangle label

Node number

DIMENSION (1)   ANGLE (2)   ANGLE (3)

TRIANGLE B

TRIANGLE A DIMS
TRIANGLE B DIMS
TRIANGLE C DIMS

TRIANGLE D DIMS
TRIANGLE E DIMS
TRIANGLE F DIMS

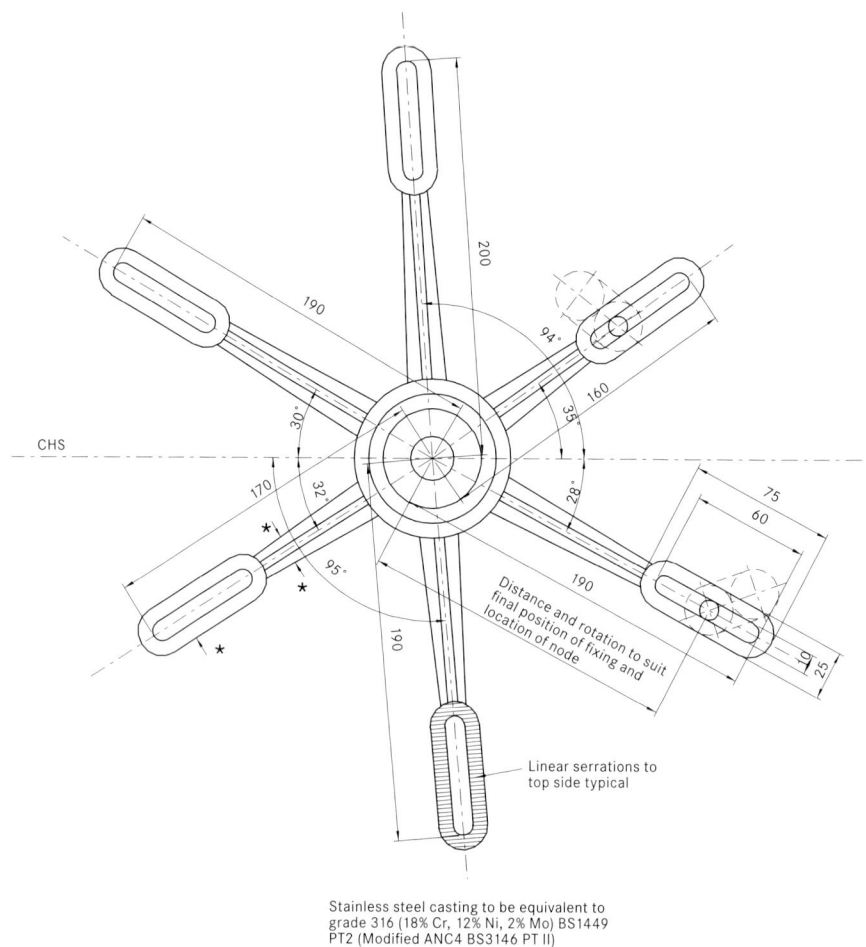

200

190

94°

160

35°

CHS

30°

170

32°

28°

95°

190

190

75

60

19

25

Distance and rotation to suit
final position of fixing and
location of node

Linear serrations to
top side typical

Stainless steel casting to be equivalent to
grade 316 (18% Cr, 12% Ni, 2% Mo) BS1449
PT2 (Modified ANC4 BS3146 PT II)

★ - Indicates finished surface to be polished

Drawing of spider fixing

So all you need to know is that each knuckle has a radius around it, the hole will fall somewhere within that radius, and that you can accommodate it.

Right. And that accommodation comes from three devices: the length of the arm, the angle—this is a parametric model—and also the size of the slot. The Eureka! moment, if there is such a thing, is undoubtedly the creation of this device. The idea was conceived collaboratively with colleagues, and it came about through playing with a physical cardboard model. There is a real combination of engineering and product design, involving a material issue, a technological issue about glass, issues surrounding the choice of casting material, and also design considerations such as the antislip serrations.

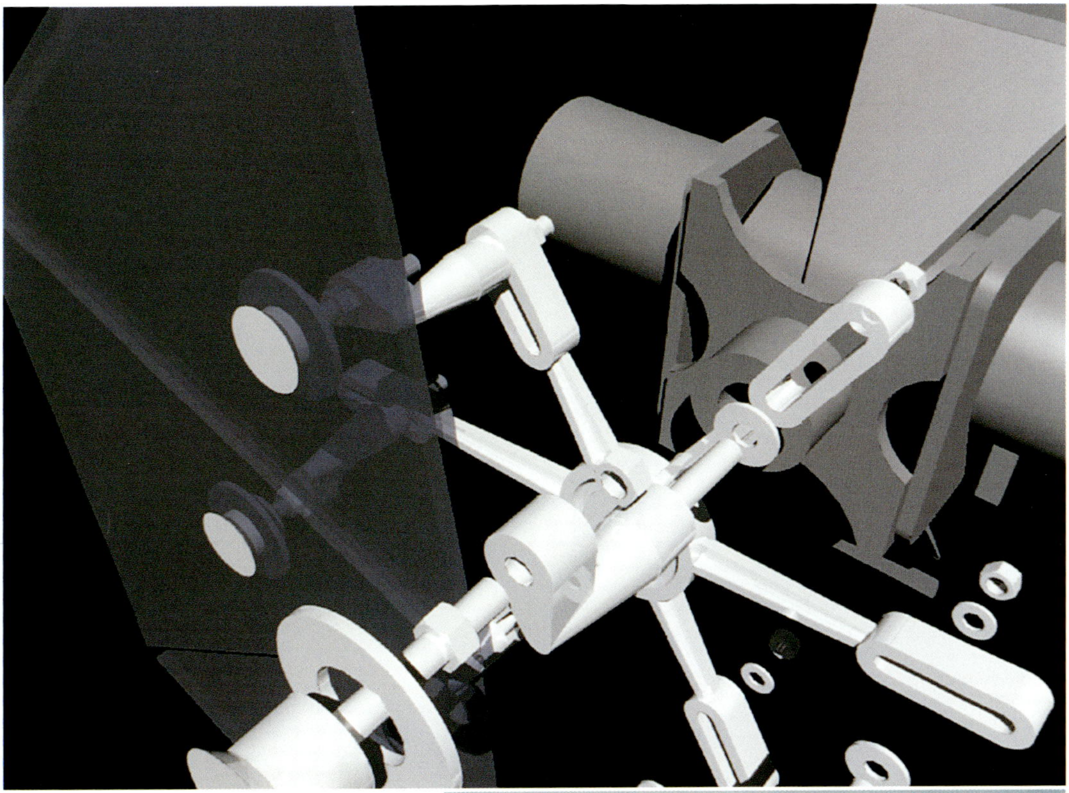

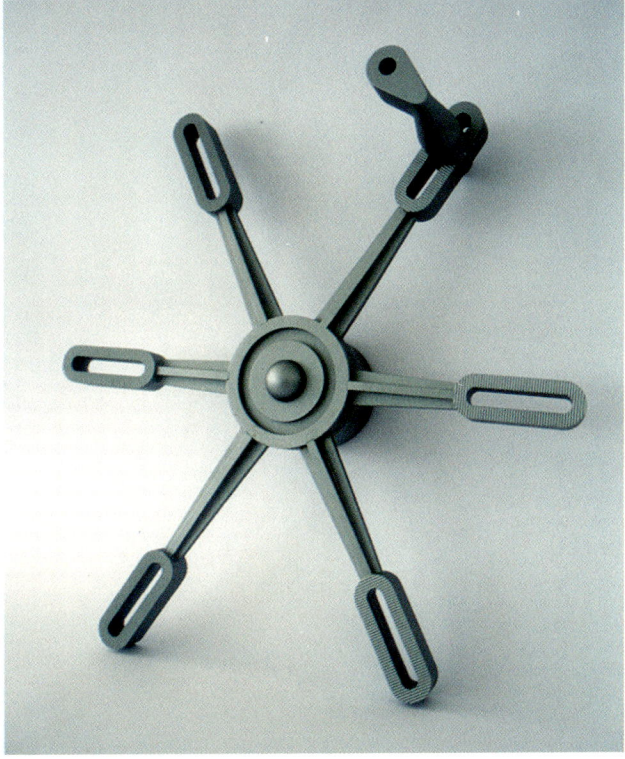

Visualisation of connection

Completed wall

Produced by Optima and intended to be erected in shops and public concourses, the Hutchinson 3G "igloo" is a recognizable image and point-of-sale kiosk for mobile phone and IT services. AKT analyzed FOA's complex design for a three to four metre diameter, self-supporting glass structure by using a finite element computer model to define the overall geometry. Full-scale mock-ups tested the solution for connections. The computer model examined load patterning across the hexagonal glass panelled surface, and showed that each joint was different. A flexible hinge was devised to cope with the variety of conditions.

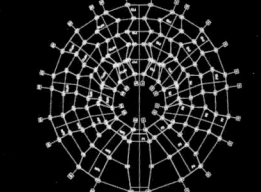

PODS, HEXAGONAL GLASS AND GLASS JOINT HUTCHINSON 3G

FOREIGN OFFICE ARCHITECTS
LONDON, 2002
CLIENT: HUTCHISON 3G

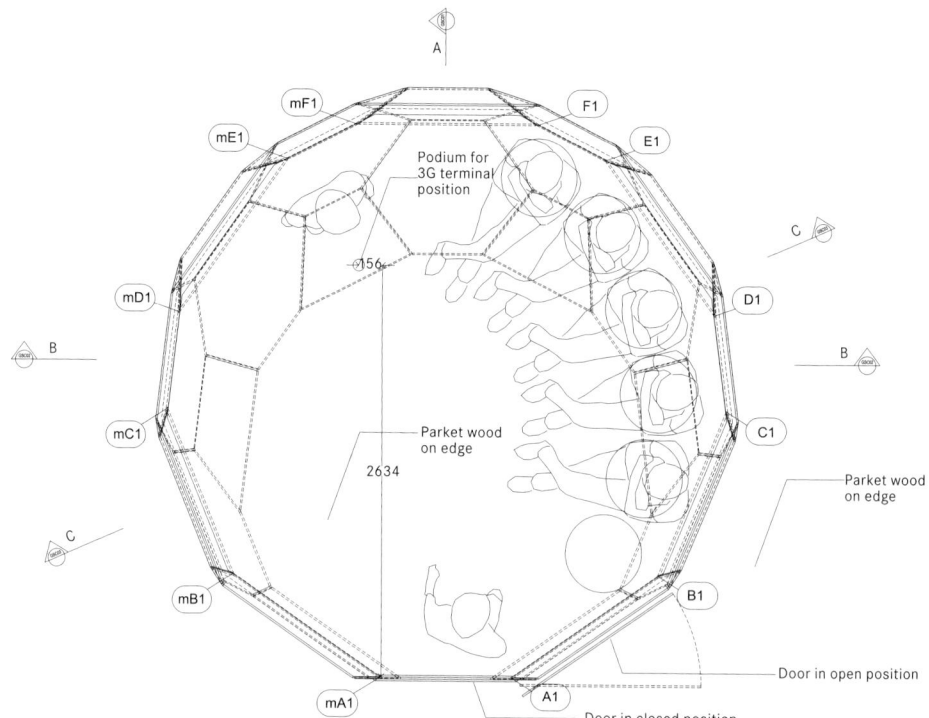

Labels in diagram:
A
mF1
F1
mE1
E1
Podium for
3G terminal
position
756
C
mD1
D1
B
B
Parket wood
on edge
2634
mC1
C1
Parket wood
on edge
C
mB1
B1
Door in open position
mA1
A1
Door in closed position

FOA's general setting out of closed pod

FOA developed a project based on a hexagonal tile pattern for this 'prototypical' idea. When we were chosen as engineers, it allowed us to further develop the work that had been undertaken on the Southwark glass wall. In this case, the challenge set by the architects was to develop varying sizes of a hexagonal panel that could fold and unfold to provide curved and planed internal walls. On this occasion we tested joints at the node and developed solutions based on this, but found these to be inflexible and less attractive. The 'prototype' solution was based on moving away from the 'node'. The new 'hinge' type joint then becomes a universal fixing, releasing the architects to develop hexagonal patterns of varying geometry. Although the 'prototype' as a product was to be used on many sites, the first site to be tested was in central London.

We had been examining tube structures, and contemplated accommodating the new spaces inside a tube. We would then take a regular existing office building and turn it into a network. It would be like walking inside a computer, through the cables, and seeing all these different things around. But the team was then given an existing building with various

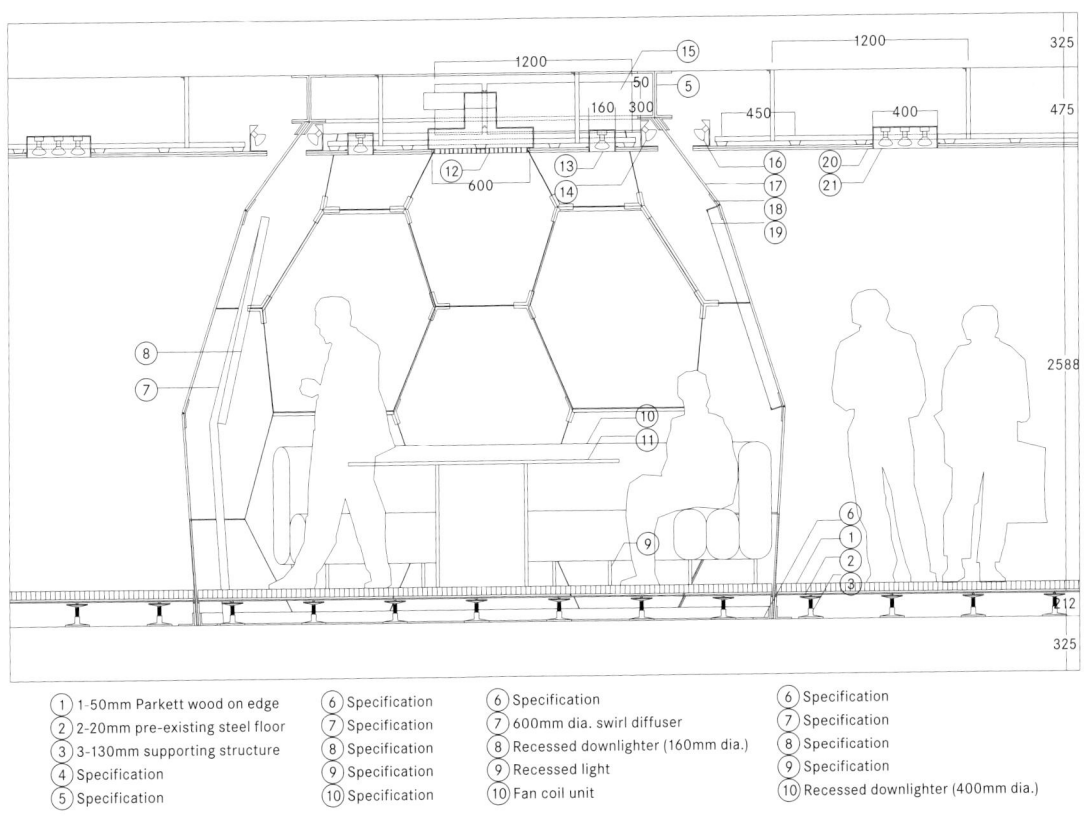

(1) 1-50mm Parkett wood on edge	(6) Specification	(6) Specification	(6) Specification
(2) 2-20mm pre-existing steel floor	(7) Specification	(7) 600mm dia. swirl diffuser	(7) Specification
(3) 3-130mm supporting structure	(8) Specification	(8) Recessed downlighter (160mm dia.)	(8) Specification
(4) Specification	(9) Specification	(9) Recessed light	(9) Specification
(5) Specification	(10) Specification	(10) Fan coil unit	(10) Recessed downlighter (400mm dia.)

floor plans and internal layouts. The client wanted a flexible design that could evolve into different spatial entities such as the entrance, the pod or an unfolded wall. It also needed to have some relationship to networks encompassing a hexagonal diagram. The brief also asked that the design be flexible so that it could be adapted to a new building. Our proposal took something from the original brief by using a hexagonal pattern instead of a triangulated one. One of the problems with using triangles is that it generates more waste of materials. Hexagons therefore allow more use out of the glass sheets as, geometrically, they fit within a rectangular shape much more efficiently than triangles.

As we developed the design, it became clear that conditions within the existing building were such that we would have to vary the dimension of the hexagons in order to occupy different zones. The 'prototypical' tile would need to additionally join and be capable of being unfolded and folded to deal with a variety of internal conditions ranging from varying floor to ceiling heights, to sometimes creating a closed private pod, whilst at the other extreme, constructing a flat wall between the two spaces.

First prototype

Economically, what would be the worst case?

Curving it is the worst scenario as we have to maintain integrity of the joint and deal with a variety of angles, whereas completely unfolding it is the best. So we examined these polarised ranges in terms of geometric change, and we started to plot and draw them. The node at which the hexagons come together is slightly different from that of six triangles meeting (as in Southwark). The first attempt was to develop a 'node' where three hexagons meet out of fine machining aluminium blocks. Some tests were undertaken, and loadings were fit tested by the team. Whilst this provided a minimal joint, load tests showed that the stresses in such a joint were easily manageable. The conclusion we came to was that 'fit' on site in a variety of conditions would be too precise and lead to problems unless a repetitive tile size and geometry were chosen. The conclusion that FOA came to was that the varying hexagon was more important in the design than the fineness of proportion of the joint.

We agreed to make the hexagons different sizes, move the junction away from the node, and make a very simple joint where only two glass tiles are meeting each other, like a door hinge. Structurally it has the potential for collapsing, but it has a very simple locking mechanism built into it, and once assembled, it can be locked with a 'blind bolt' to prevent it from folding on itself. The assembly is flexible in the unfolded shape and very stable for a closed shape. A full scale prototype of the folded and unfolded forms was tested by the team for acoustics and graphics, but the project did not go further as the clients' strategy changed. The work done to reach this stage of constructing and testing the full scale prototype was very valuable from everyone's perspective, not least for us to again develop analytical skills that make good use of new tools. The work done during the design of the Jubilee Line glass wall project had been very relevant. The shift from the 'universal joint' with less degrees of freedom that comes out of shifting the attention away from the point to three separate 'lines' was the step forward.

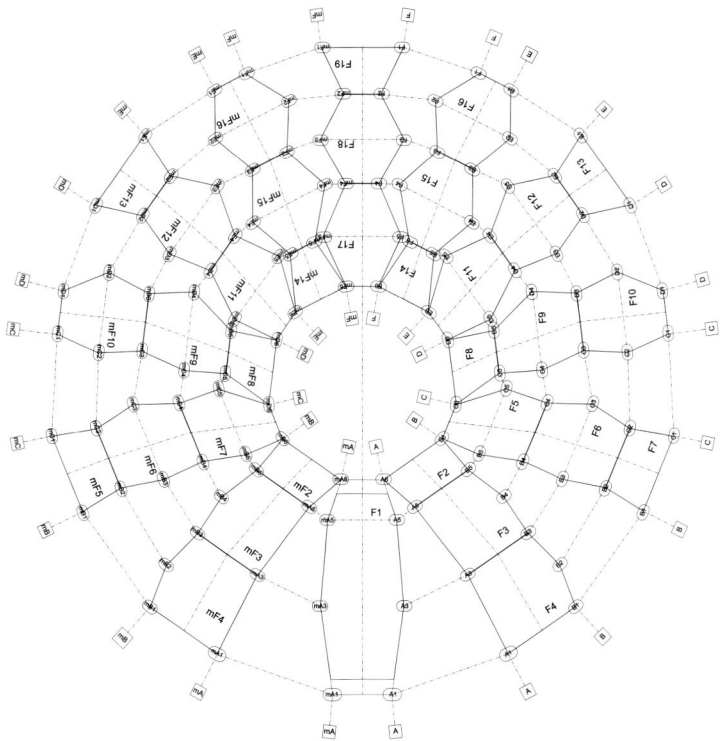

Unfolded geometry

Hexagonal glass pieces

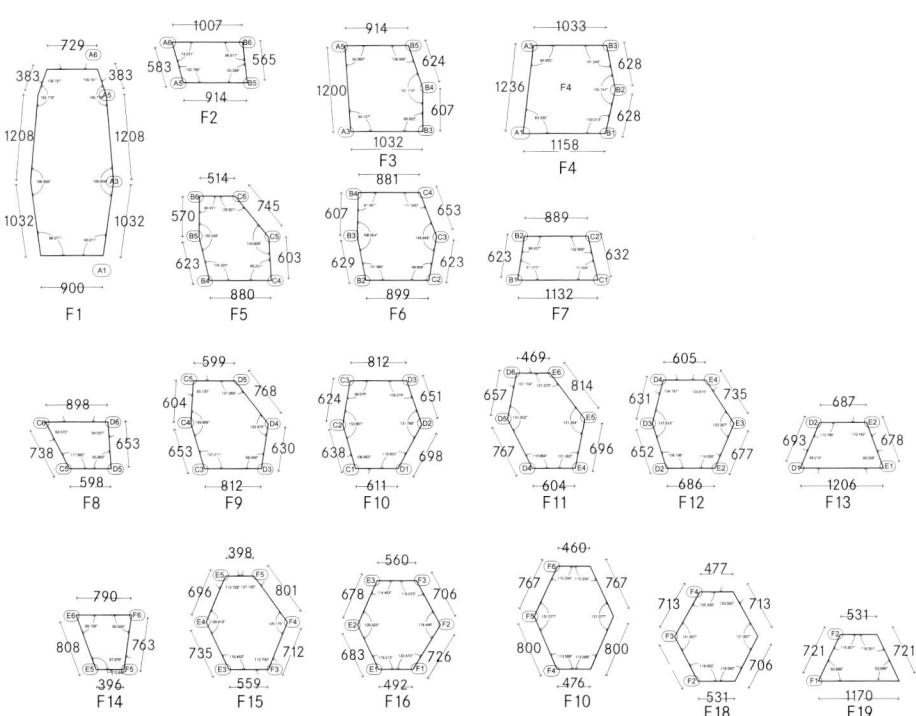

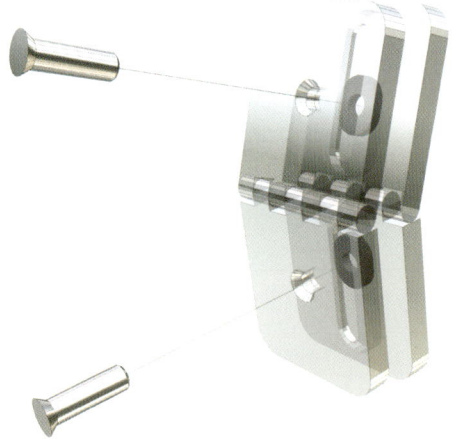

Simple hinge (adopted)

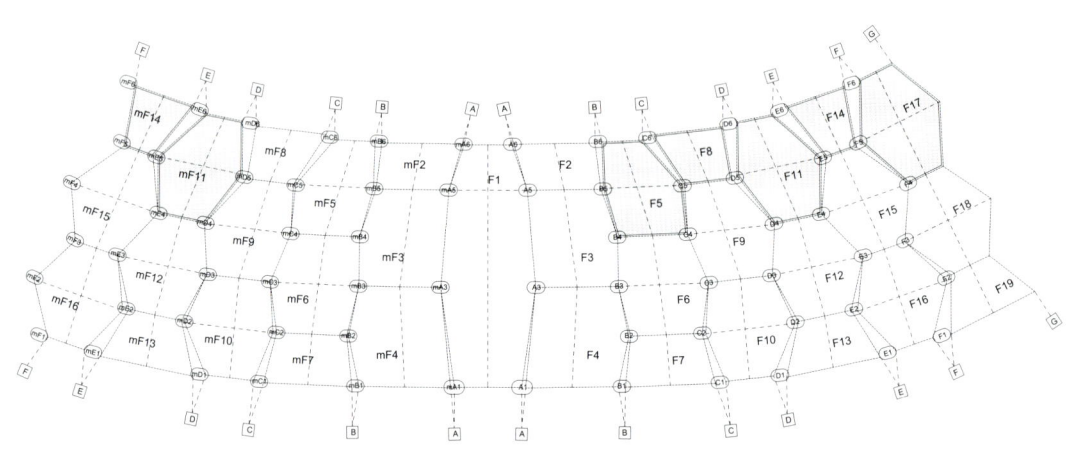

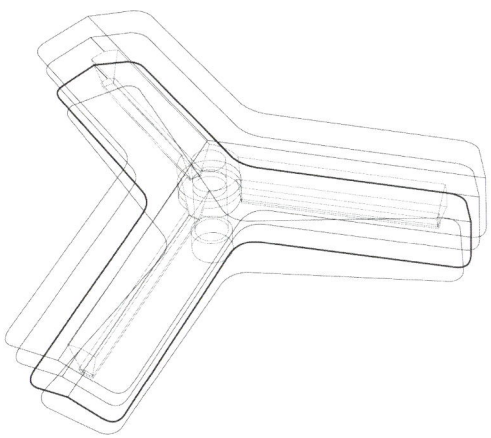

Corner node and its variation (tested but not adopted)

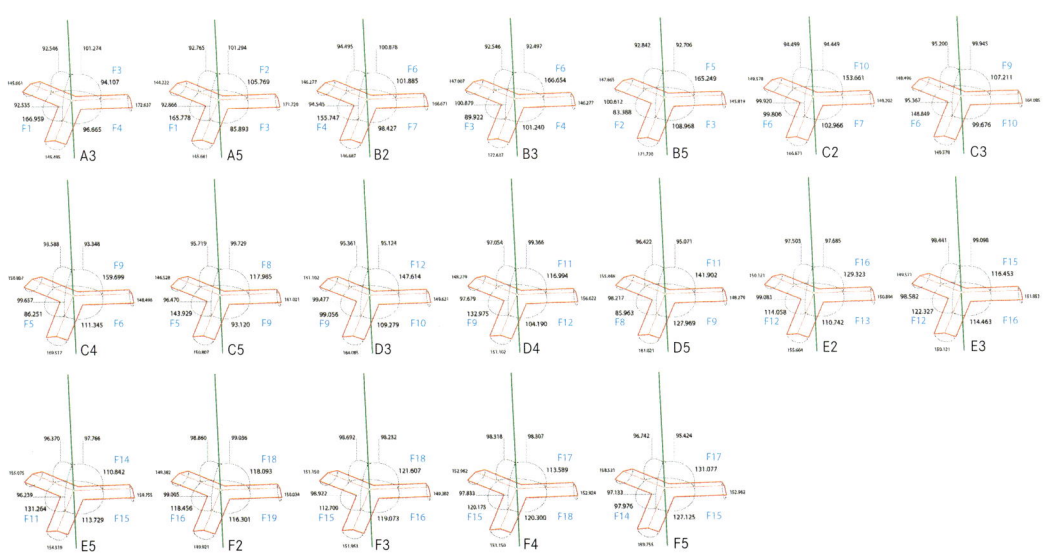

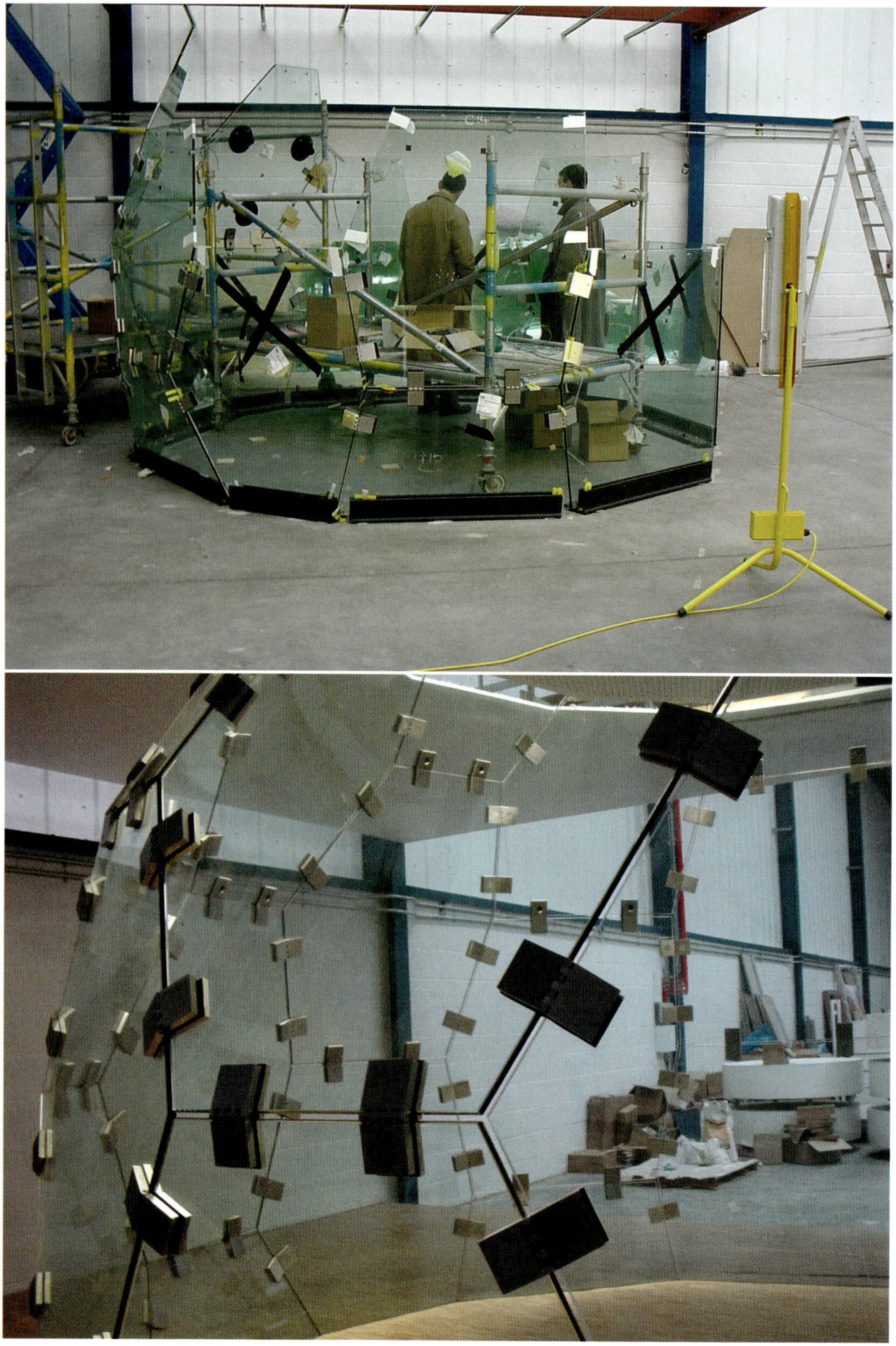

Assembly in existing building

Hinge details

This project was about simplifying the problem, as opposed to the solution at Southwark. But you did try it the more complicated way first?

Yes, but it didn't work. We had learned the fabrication methods that were used were precise but there was a greater risk of lack of fit in this case as the 'prototype' was likely to be used in a variety of situations (unlike the given site for the Jubilee Line wall). Visibility was also an issue. The structure in the train station works because it was a backing for something that was never seen, so less attention was necessary. But in the case of 3G, the project was two-sided; it would always be visible from both sides.

We even tried plastics for the node. We looked at advanced composite plastics so that it could become transparent and so on, but such complexity was unwarranted. So instead we simplified it and developed the hinge. The hinge looked great in a drawing, but when we made the physical prototype, where we could see through several layers at the same time, the aesthetics were different as the hinges didn't define the hexagons in a 'closed' pod wall. In the more translucent case this issue was less concerning. What we were able to prove was that many shapes could be assembled at speed with this approach.

What did you learn from the project?

We developed a better understanding of the glass 'clamping' action that is often not easy to get from manufacturers and have gone on to use this for later projects. It's another reason why we conducted physical tests, because of those things that are impossible to calculate or find in codes and standards. The combination of physical testing and applications such as this is sometimes the only way to innovate in materials like glass.

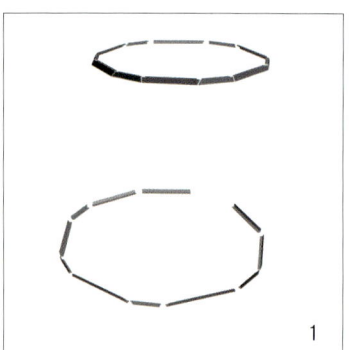

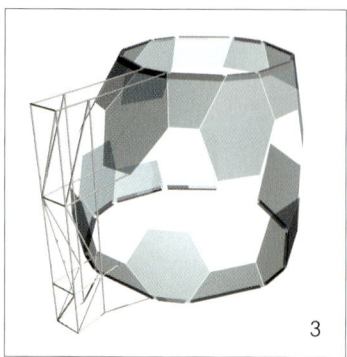
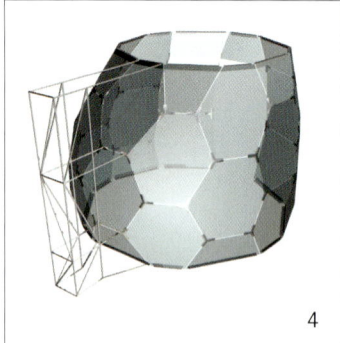

Construction sequence

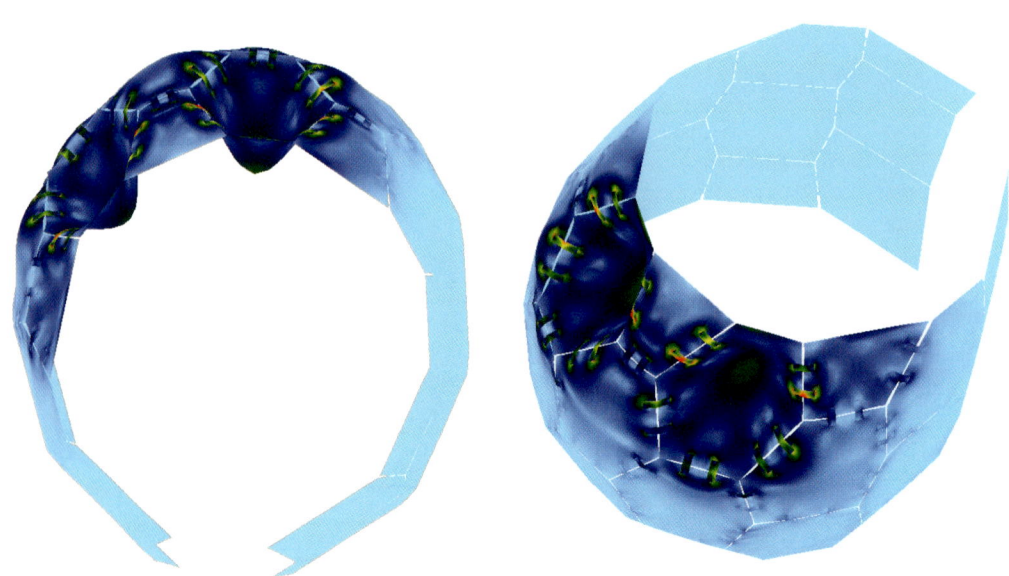

Finite element analysis in various temporary and permanent sequences

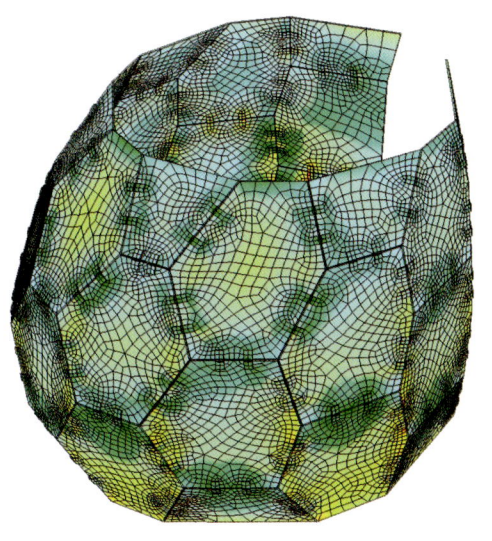
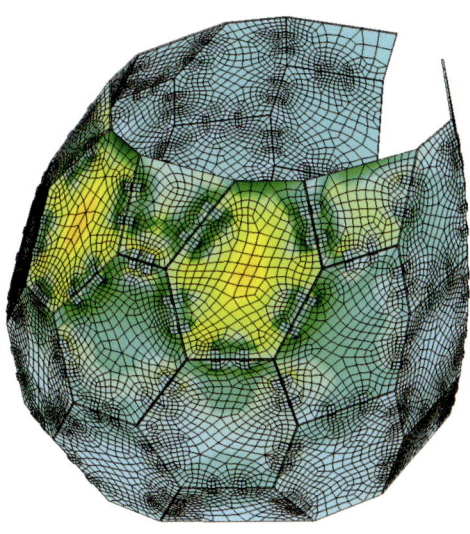
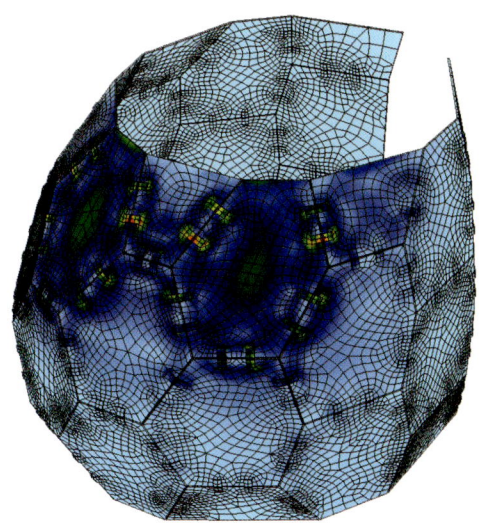
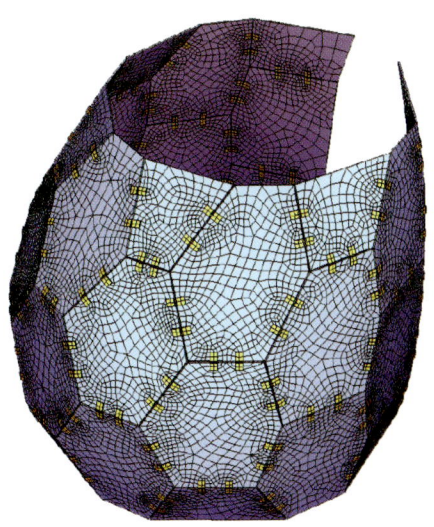

Designed by Zaha Hadid Architects, the train station is located three kilometres north of Naples on a new high-speed line. Its form derives from the lava flows typical of the Napolitan campagna and its dominating feature, Mount Vesuvius. In a collaborative role, AKT understands and develops the design's mathematical qualities so that the metaphorical ones can be reinforced. An S-shaped bridge above the tracks houses the ticket hall and provides entry to the platforms. The 500-metre long by 45-metre wide station is a hybrid structure. AKT used concrete, moulded into deep T-beams, up to the second concourse level to span the twenty-eight metres across the track, and to cope with active seismic conditions. Steel portal frames shape the shed-like, sky-lit spaces above.

ASYMMETRICAL BRIDGE
NAPLES HIGH SPEED TRAIN STATION

ZAHA HADID ARCHITECTS
NAPLES, ITALY, 2003-2008
CLIENT: TVA

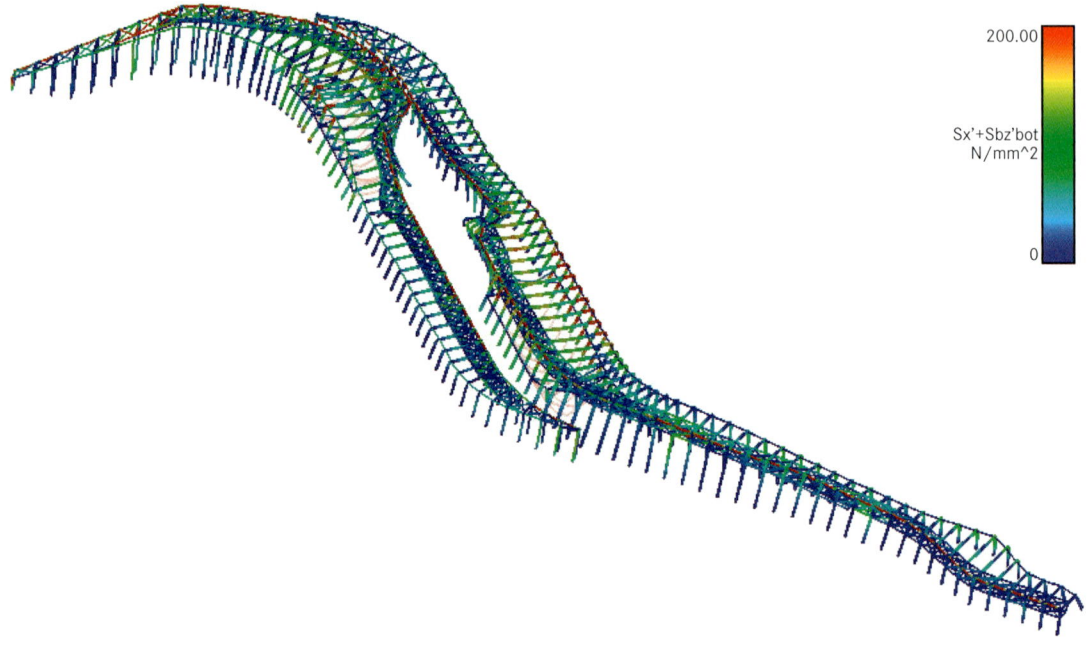

200.00

Sx'+Sbz'bot
N/mm^2

0

Static Case: 1.2 + 1.2 + 1.2 W1 Defl. (mm)

We have been collaborating with Zaha Hadid Architects for some time and have developed our thinking through some of the questions they set us, so when we start a competition with them we are very quick to converge to a solution. In this case, early discussion came to the conclusion that we wanted to put the railway station over the tracks as a way to connect the two parts of the city. This was a positive strategy from the city viewpoint as well as from a programmatic viewpoint, as we would be building over the railway (airspace), connecting two sides of the city and freeing valuable footprints on either side of the railway.

Zaha Hadid's office created a chicane, an S-shape, with the plan geometry derived from fixed entry and exit points. But that poses a technical problem in a seismic area like Naples, as any asymmetry in the mass, or disproportionate shape, exaggerates the impact of earthquakes on such structures. The dynamic analysis was very much against the symmetrical organisation you would normally encourage, but the benefits of the form outweighed the technical challenges, which hence gave us a new opportunity to develop asymmetric structures in such zones.

As the analysis and design were likely to go beyond current codes of practice in this area, we collaborated with Imperial College, London and brought in Professor Ahmed Elghazouli, who is one of the professors in earthquakes and who has worked with us on other projects. There are certain situations in which you need to collaborate with the best experts in order to push the boundary further. Often such collaboration with experts releases both parties into a new direction whilst unleashing years of potential on each side.

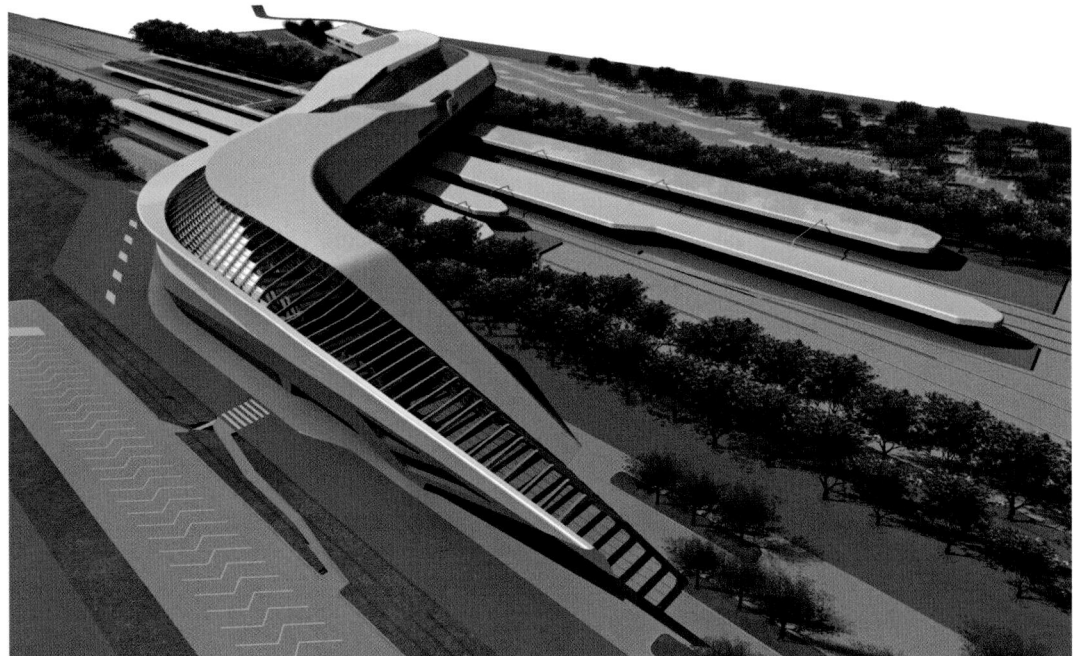

Zaha Hadid's competetion visualisation

Similar to the Phæno Centre, this is a seamless project, with the architecture and the engineering being conceived together from the inception at the competition phase. The way it works is that we developed the system so that the main central portion is very straight, knowing that the infrastructure might or might not be there in time. With the conventional geometry of the middle portion, we proposed a fairly simple pre-cast, pre-stressed bridge deck acting as the floor. Everything from the roof transfers down to the deck level, which supports and transfers all the loads like a bridge, except that the supports had to be carefully located to allow the transition to ground at both ends as a curve.

So, it's a deck-bridge with "things" on top?

The structural elements are simple and of common systems in Italy, complex only from a geometric point of view. For instance, every piece of steel and cladding is different. There isn't any particularly spectacular structural feat, apart from the earthquake modelling. Imperial College worked with us to challenge the codes of practice and then to develop the structural system with us, feeding us information in terms of how we should model it, and what were common problems in this particular region in terms of earthquakes. So instead of simplifying the intent and forcing the design to conform to local practices, we challenged conventions and empirical values by bringing the general expertise of Imperial College to the specific design and site.

Detail showing typical braced walkway arrangement

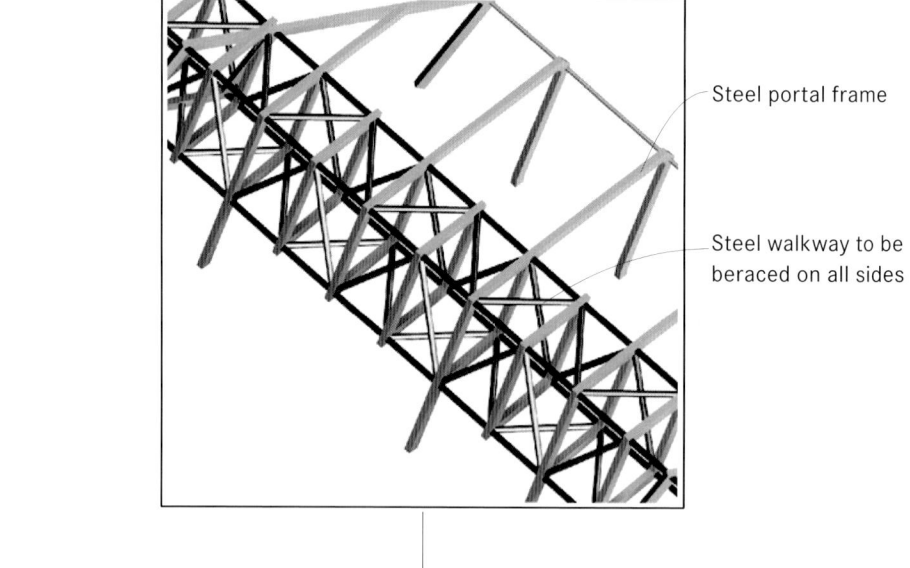

Steel portal frame

Steel walkway to be beraced on all sides

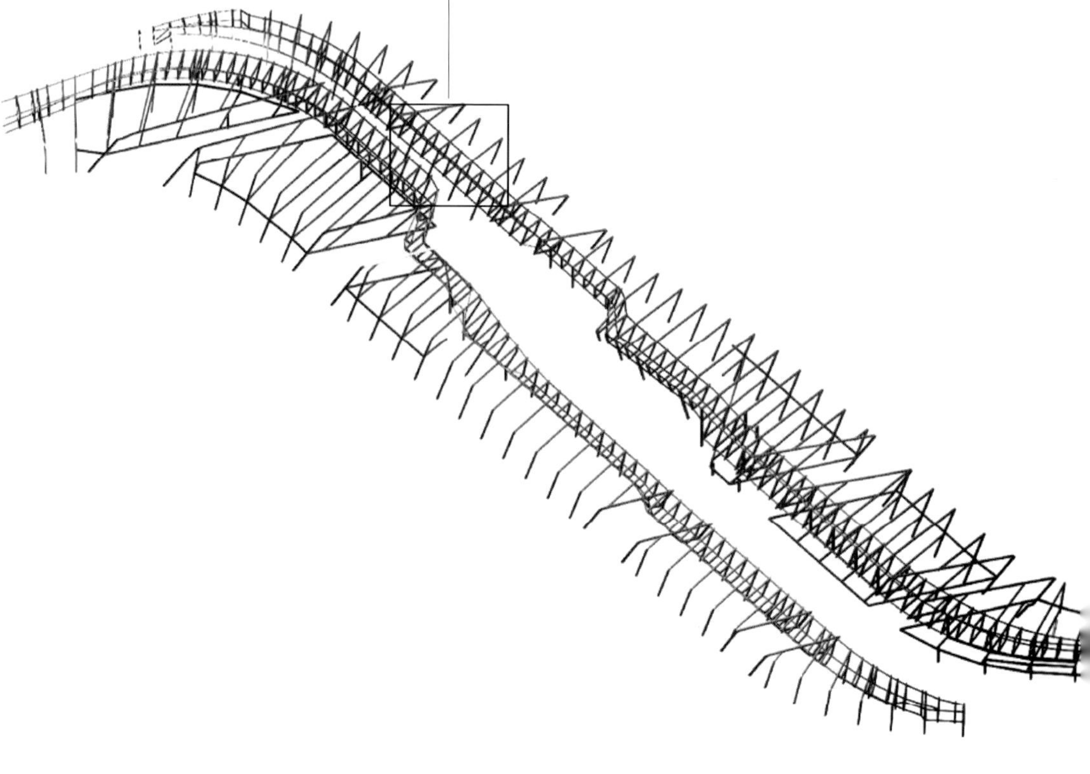

Isonometric of steel portal frames with braced walkway removed for clarity

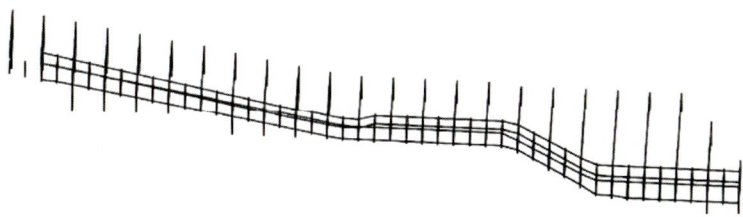

It's interesting how you didn't reduce the complexity of the problem, choosing to embrace it instead

With Zaha Hadid's office it's like that on most projects. The confidence that we developed after the Phæno project helped us to become even more comfortable with challenging certain norms. The work is usually won through design competitions where the competitors are all keen to push the boundaries and are leading offices in the contemporary scene. Thus we are all tested technically and forced to invent wherever we can, developing a certain shorthand with the offices that we work with regularly.

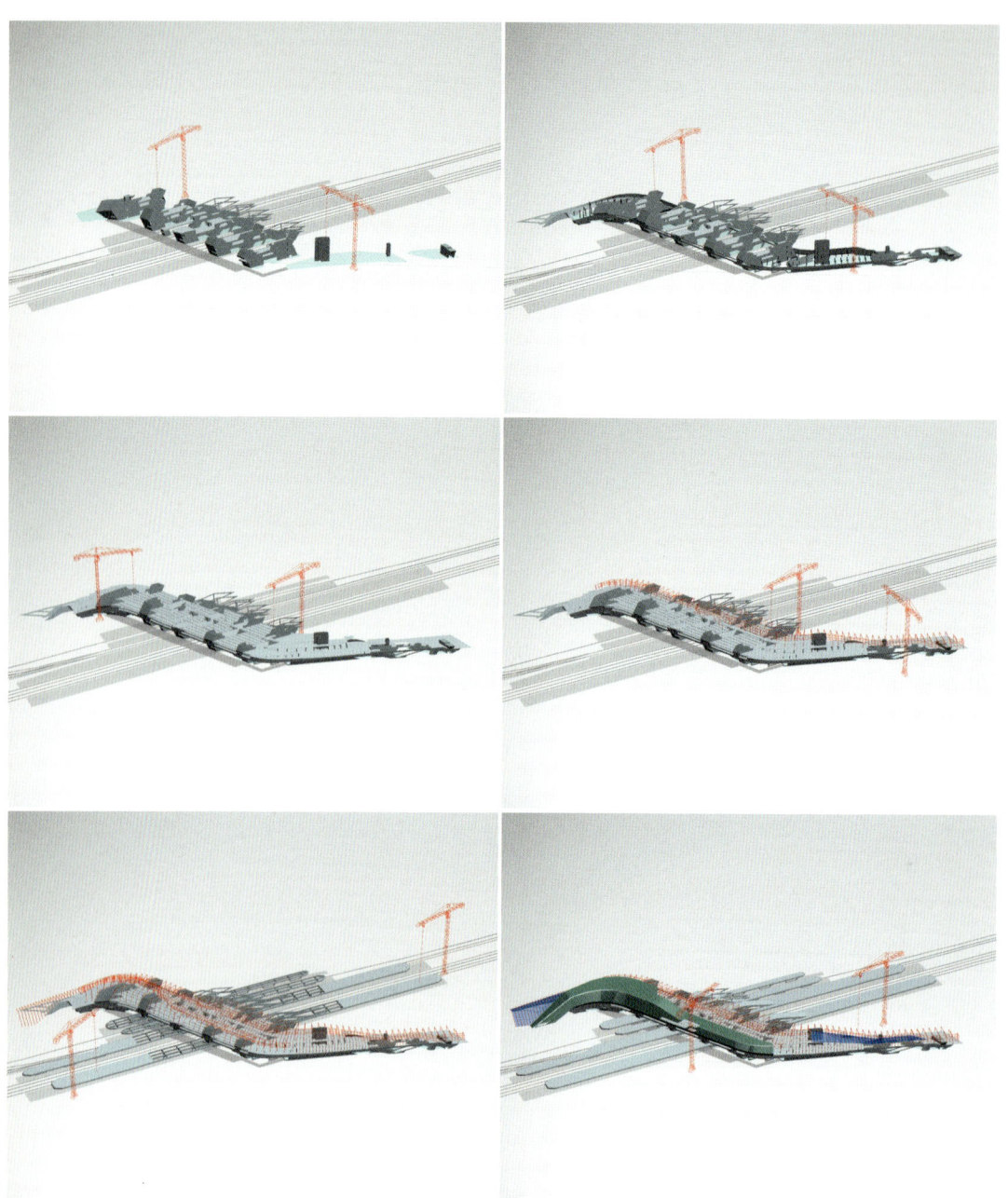

Construction sequence

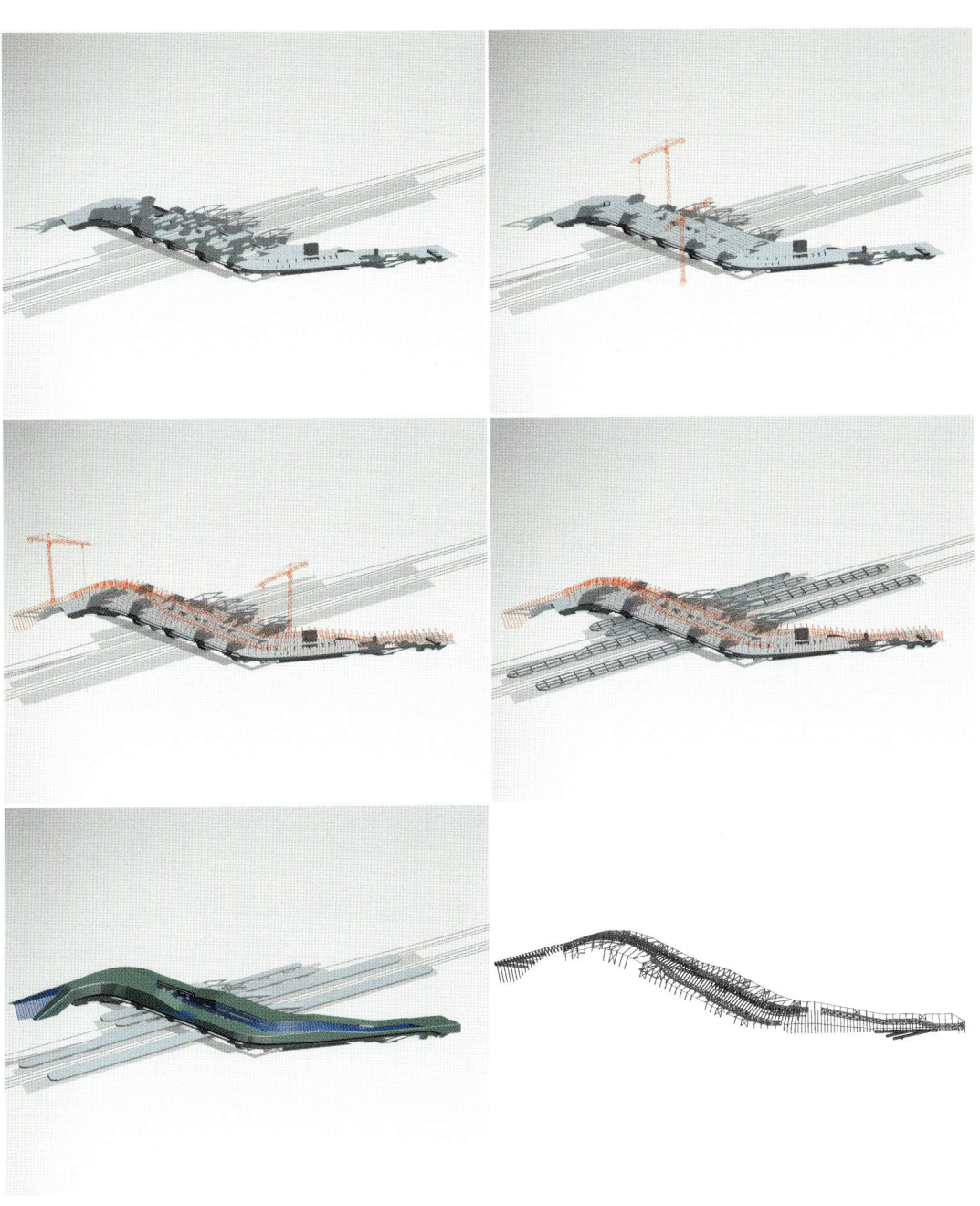

TRANS-SCALARITY (MICRO-MACRO)

Sometimes the smallest decisions in a project can have the largest effects on the overall design. Part of design engineering involves knowing at what scale(s) the key decisions lie in each project, and at what scale particular problems need to be solved. This ability for the micro to affect the macro, and vice-versa, is closely linked to the relationship between architect and engineer, as sometimes the decisions that are made at the smallest scale of design engineering are responsible for having the largest impact on the outcome of a project.

New calculating tools are now capable of producing unprecedented immediacy across scales. Zooming capacity and the precision of nested geometries in electronic engineering are binding the micro and macro scales of projects to converge seamlessly. Where just a few years ago expertise was typically differentiated by its scale of operation, electronic documentation is inextricably connecting different scales of organisation, blurring the conventional limits between the disciplines of urbanism, architecture and design engineering. The co-existence of different scales in a single tool or document allows the rethinking of the relationship between the part and the whole. The joints between parts and their reference to the system, the differentiation of the repetitive construction element, and its 'localisation', are some of the architectural possibilities that electronic documentation or numerically controlled manufacturing are opening for experimentation. The relationship between this 'productive' power (through the authority of digital technology) and human action (through the engineer's intuition) is an area that needs balance if we are to produce successful ideas in the future.

General information of the Peckham Library
see chapter PROCESS, case PODS, page 16

SLOPING COLUMNS
PECKHAM LIBRARY

ALSOP ARCHITECTS
PECKHAM, LONDON, 1999
CLIENT: LONDON BOROUGH OF SOUTHWARK

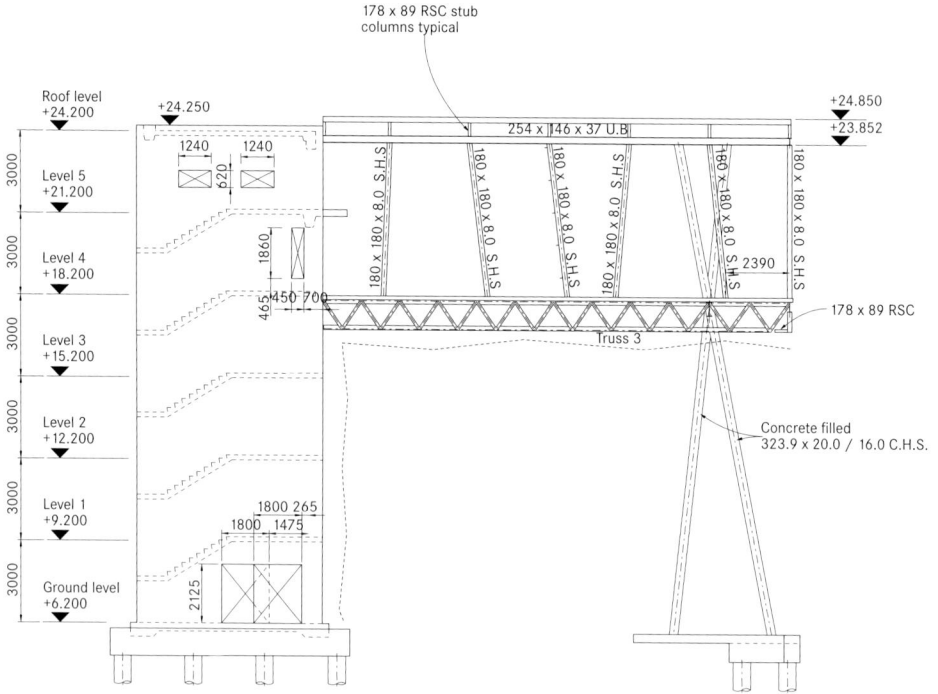

Typical sectional arrangement

When Will Alsop's office first proposed a suspended library with a column-free space underneath, we thought that his proposed cantilever was totally unfeasible, and that it wouldn't be affordable. His proposal meant that the notion of 'using columns' was to be challenged. We therefore tried to create a choreography of columns at different angles that complemented the architectural ambition, and that could simultaneously make the scheme feasible. We developed a new way of thinking about columns, thinking of them as a 'group' first and then returning to consider each one as vertical support in the conventional sense.

Introducing this 'complexity' to the process seems to make the life of the engineer worse. Although it solves a problem, does it also give the architect a space in which he has freedom to operate?

Absolutely. Instead of controlling the geometry with a pointgrid, where the designer has no room to move, we defined a 'virtual cylinder' that circumscribed the different positions of the columns at the ground floor and roof planes. We wanted to give the architect the freedom of actually using the structural arrangement of the group as an architectural idea. The resulting system seems random, but it is largely controlled: the columns pivot under the floor where they support primary trusses to land within a controlled circle at the foundation level. This variety of foundation points also allowed the manipulation of the landscape to assist in preventing vehicle impact.

Temporary tripod to reduce scaffold

Temporary tripods removed

So, the problem induced further changes in the project?

In this case, the unusual column geometry became a positive contribution to the project. The design issue that was raised in the practice was about the way of working, because we had suggested something that cost more (elementally), and had to keep costs similar to that of a vertical column. To build something fourteen metres up in the air requires temporary scaffolding; since all these leaning columns were unstable individually until the group was in place, we had to create a temporary 'tripod' to stabilise each column. Normally when building a floor, it would be supported with a scaffold; what we did in this case was to temporarily introduce two simple additional short legs to the column to create a tripod, thus dramatically reducing the use of temporary work.

Having suggested the idea as a 'whole', the remaining challenges were to find the slenderest column section for that height and connecting the plane trusses that support the floor to the columns. The slender column of 323mm diameter was achieved by using composite concrete filled circular hollow sections. It was a method used in certain parts of Europe but largely untried in the UK at that point. The connection of primary floor trusses to this column was complicated not only by the varying angles of each of the columns, but also by the desire to cantilever the horizontal structure beyond the column on all three sides of the building.

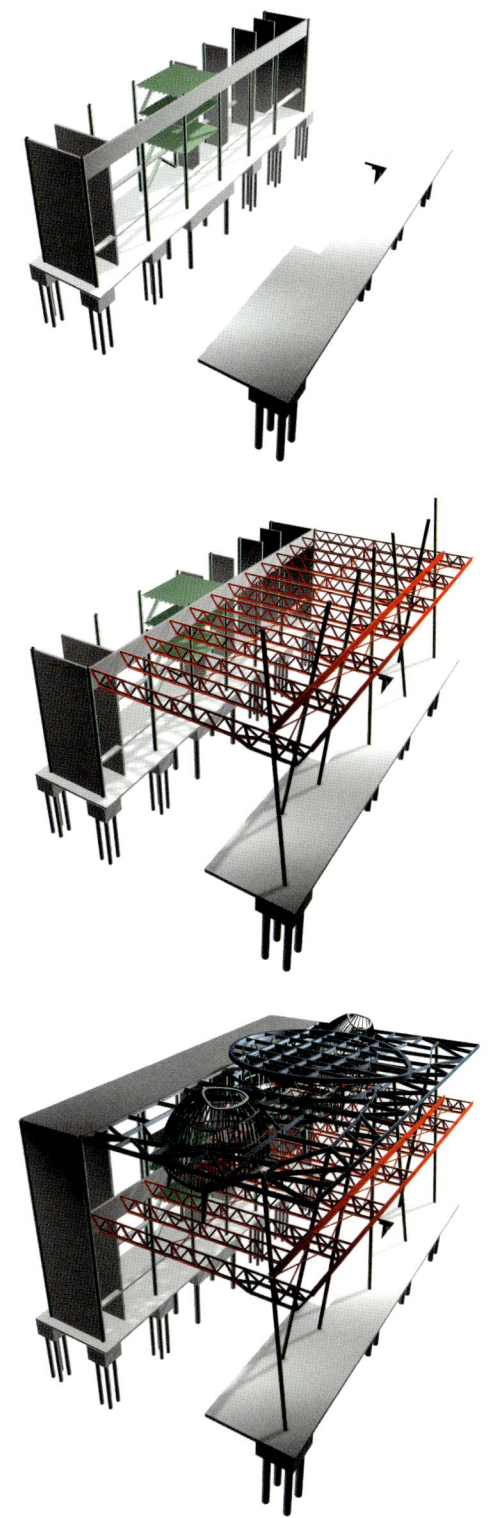

Construction sequence

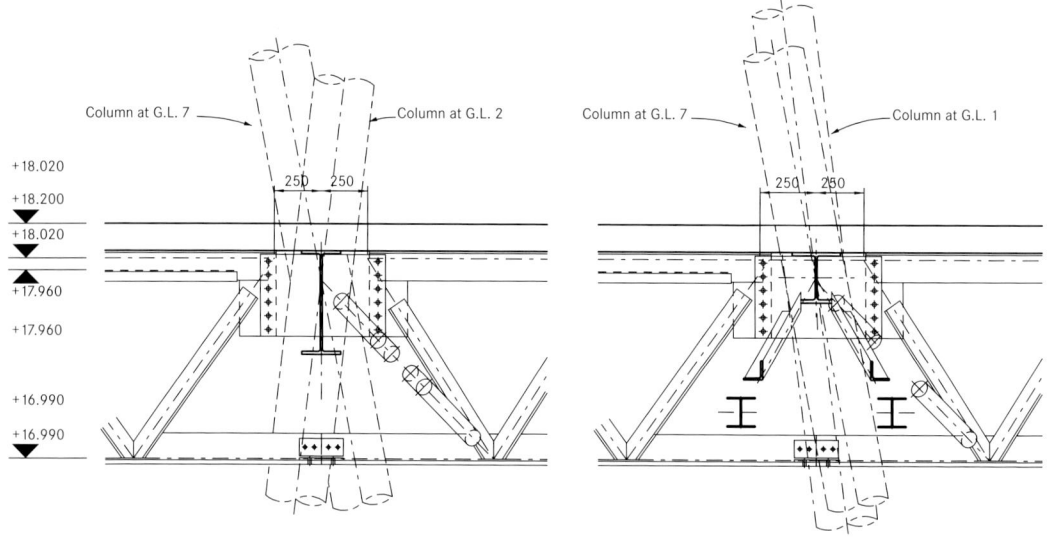

Typical elevation on truss 1
at column connection

Typical elevation on truss 1
at column and truss 4 connection

How does the connection actually work?

The plane truss bifurcates around the column. The column tube has a slot cut in it with
a plate going through the column, and this plate allows for connections on both sides to
the top plane of the truss. The actual connection is complicated by the fact that the tubes
are concrete-filled to provide additional strength so the plates had to be fabricated prior
to pouring concrete through the steel section and compacting the concrete. The bottom
flange of the truss is not connected. This arrangement deals well with the cantilevering
effect of the building. Unlike the main span, where the bottom of the truss is in tension,
the cantilever reverses the action so the bifurcated tops are in tension. The forces from the
cantilever at the front of the building go back and up the truss because of that bifurcation,
and the two ends of the building work in a similar fashion. This connection became the key
to the 'whole' system and the details had to ensure that not only a variety of column angles
are accepted, but that the assembly is simple and stresses developed are locally tolerable.

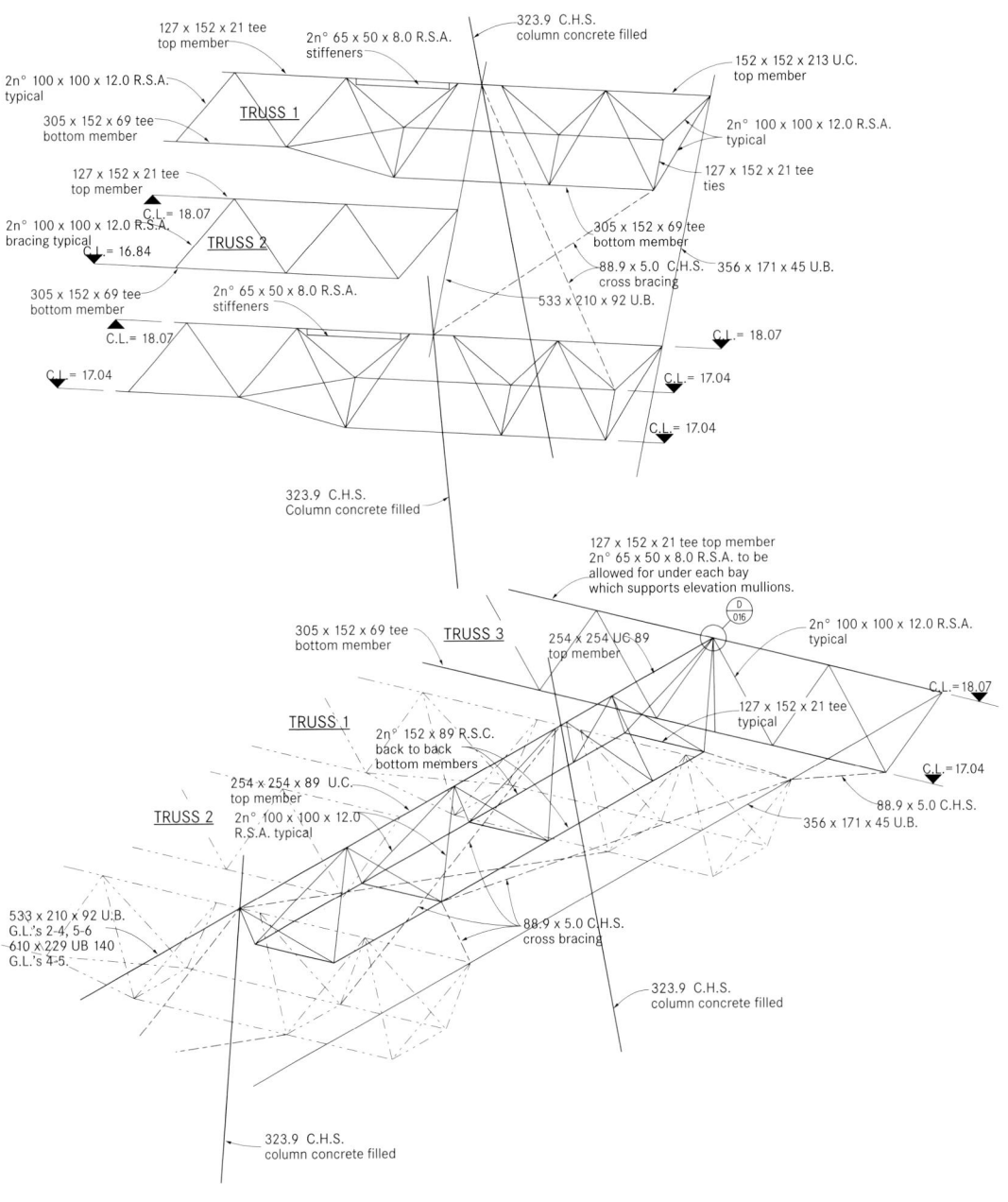

127 x 152 x 21 tee
top member

2n° 65 x 50 x 8.0 R.S.A.
stiffeners

323.9 C.H.S.
column concrete filled

2n° 100 x 100 x 12.0 R.S.A.
typical

152 x 152 x 213 U.C.
top member

TRUSS 1

305 x 152 x 69 tee
bottom member

2n° 100 x 100 x 12.0 R.S.A.
typical

127 x 152 x 21 tee
top member

127 x 152 x 21 tee
ties

C.L.= 18.07

2n° 100 x 100 x 12.0 R.S.A.
bracing typical

TRUSS 2

C.L.= 16.84

305 x 152 x 69 tee
bottom member

305 x 152 x 69 tee
bottom member

88.9 x 5.0 C.H.S.
cross bracing

356 x 171 x 45 U.B.

2n° 65 x 50 x 8.0 R.S.A.
stiffeners

533 x 210 x 92 U.B.

C.L.= 18.07

C.L.= 17.04

C.L.= 18.07

C.L.= 17.04

C.L.= 17.04

323.9 C.H.S.
Column concrete filled

127 x 152 x 21 tee top member
2n° 65 x 50 x 8.0 R.S.A. to be
allowed for under each bay
which supports elevation mullions.

305 x 152 x 69 tee
bottom member

TRUSS 3

254 x 254 UC 89
top member

D
016

2n° 100 x 100 x 12.0 R.S.A.
typical

C.L.= 18.07

TRUSS 1

2n° 152 x 89 R.S.C.
back to back
bottom members

127 x 152 x 21 tee
typical

C.L.= 17.04

254 x 254 x 89 U.C.
top member

TRUSS 2

2n° 100 x 100 x 12.0
R.S.A. typical

88.9 x 5.0 C.H.S.
cross bracing

356 x 171 x 45 U.B.

533 x 210 x 92 U.B.
G.L.'s 2-4, 5-6
610 x 229 UB 140
G.L.'s 4-5

88.9 x 5.0 C.H.S.
cross bracing

323.9 C.H.S.
column concrete filled

323.9 C.H.S.
column concrete filled

Isonometric view of truss's 3 and 4

Primary steelwork in place with little temporary works

Foreign Office Architects' design for a John Lewis store and cinema at Hammerson's Highcross Quarter shopping centre in Leicester dramatically synthesises the complexities of site and programme. Much of the drama is supported by effective engineering strategies to provide various spans for the store atrium, auditorium, loading bay and to minimise vibration and sound transfer between the different parts. Entered by ascending an amphitheatre at one corner, the store's interior is a cornucopia of goods, sensations and structural gymnastics, with several floors of seemingly structureless glass walkways rising through an atrium. The corner itself stretches the potential of the reinforced concrete frame. Its top-hung cladding reflects Leicester's and John Lewis' shared history in haberdashery, with two layers of panels of fritted glass to suggest lacework. Other innovative structural elements include the foundations. These are woven between archaeology and support large transfer structures under the cinema. Inside this cantilever is a four storey high cinema over the service yard. All this was achieved whilst balancing cost against flexibility. A 36 metre footbridge designed to maximise transparency, and to minimise structure, links the store to a new multi-storey car park, crossing quietly over the six-lane ring road below.

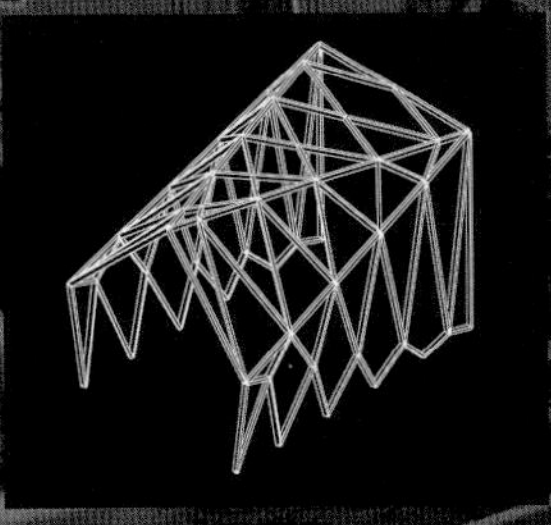

GLASS SKYBRIDGE, CORNER JOHN LEWIS LEICESTER

FOREIGN OFFICE ARCHITECTS
LEICESTER, UK, 2008
CLIENT: JOHN LEWIS PARTNERSHIP

FOA's visualisations

The project began as an invited competition where FOA were competing with several offices. FOA invited us to support their entry although neither office had worked for this client nor completed a new department store by then. When we started to talk about the brief, the discussion centred on how to insert two very large blank or opaque programmes such as the department store and the cinema, within the constraints of a given masterplan.

We discussed strategies, what could be welcome in middle England, and the 'local' specific conditions in the town. It's an unusual city in the UK because of its location and its history which has attracted a large Asian population over the last twenty years. FOA were conducting a lot of research in-house about transparency, including studies on transparent systems of façades for department stores. In essence, they were going back to a more historical approach. (These stores have only become opaque in the last thirty years).

A major contradiction in this project was that one part was completely resolved; we knew that the tenant, John Lewis, was onboard and was part of the selection and competition process, whilst the adjacent cinema block lacked an exact programme description at the competition stage. All that was specified was that it would be an opaque box, which FOA decided to define as a shiny box, while for the department store they chose to use 'lace' as a pattern and create a semi-transparent envelope, as a fully transparent building would be impractical and environmentally less attractive. Initial versions of this idea took the form of an external veil made of high strength concrete shaped as lace. Another material suggested was stainless steel that could cover the glass on the outside and achieve a certain degree of opacity, whilst also providing an environmental buffer (shading, etc.). The base of the building is formed with 'dark' precast concrete panels as it is close to the ground and needs a different sort of robustness.

Corner view

Another important structural intervention is the footbridge that links the main car park to the shopping centre. FOA wanted to make this part of the main shopping experience rather than an afterthought This meant making a high quality transparent bridge that would perform structurally and draw attention to the Anchor store as soon as one entered it from the car park. The competition was won with these ideas and few of the initial ideas have changed; the lace is now printed onto the glass for budget reasons and as a result of a rigorous development of the design. The outer skin is a single layer and then there are two interior layers, with a one-metre deep cavity in between the inner and outer skins.

FOA wanted the structure of the façade to disappear in order to disrupt the lace patterned façade as little as possible and to make the façade look as seamless as possible. We looked at various options, but it soon became clear that the external façade needed to be hung from the top of the building to enable us to engineer out all movement joints within the glazing, thus providing a virtually uninterrupted glass façade. (A floor level supported façade, or one hung from each floor level, would need wide movement joints at each floor breaking up the envelope and interrupting the glazing pattern). The hangers are arranged at 2.4m centres around the perimeter of the John Lewis building, but only every other hanger provides vertical support to the glazing. Paired glass panels are held together at the base by a thin steel cable hidden behind the panel joints providing a strut and tie system (the glass panels act as struts). This allows the glass to span naturally 4.8m between the vertical support hangers. All hangers provide restraint against horizontal loads.

The suspended facade with minimal joints

The superstructure frame was designed with long cantilevers around the perimeter of the building, therefore eliminating columns from the elevations. This provided its own challenge as the frame is in reinforced concrete and has complex deflection behaviour. Therefore, to achieve the seamless façade FOA wanted we had to design the superstructure and façade to work efficiently and elegantly together to avoid overdesign of one or the other.

By using this approach, we effectively 'flattened out' the deflection profile of the supporting roof slab, thereby minimising the differential movement between glass panels without having to stiffen the superstructure frame unnecessarily. This allows the silicon joints between the glazing panels to be less than half the width than if supported on each hanger, and therefore match the join widths throughout the façade. The only movement joints are at the bottom of the glazed façade where it meets the black concrete façade panels and bridges.

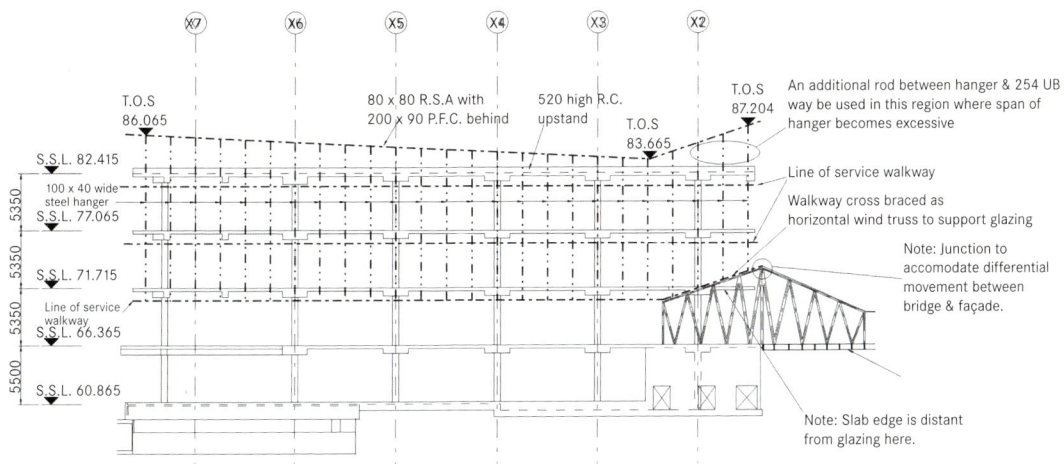

Building cantilever support to bridge

The internal glazing is supported at each floor level, and the movement joints at these levels are hidden in the change between the clear glazing and spandrel panels. This meant that the building could be made watertight quickly while allowing the more complex external hung façade installation to continue, without risk to the critical path operations inside the building.

The bridge was technically quite difficult, because it had to span 36m over a major highway with a fixed height clearance that was too restrictive. The solution was to cantilever the department store structure and have the bridge begin eight metres later, reducing the clear span from 36m to 28m. This allowed us to retain the whole idea from the competition stage, and utilise the main cantilever to act as the main entrance for the department store. In addition, an almost transparent wall and roof were also achieved through using structural glass. One of the biggest changeables was that the bridge tends to overheat due to all the glass, and careful detailing was necessary to deal with expansion.

For an engineer, it was interesting to walk on the bridge before the glass was installed, as the dynamics are finely tuned and the structure can be excited. It's only after the glass is installed that this provides the dampening of the bridge.

The desire to read the glazed pattern as a 'whole' by minimising the joints and utilising the pattern has proven to be very successful. One of the most enjoyable aspects of that result was that the procurement route allowed ourselves and the architects to work closely with the manufacturers Seele in order to achieve the same goals, without necessarily resorting back to known joint types and solutions.

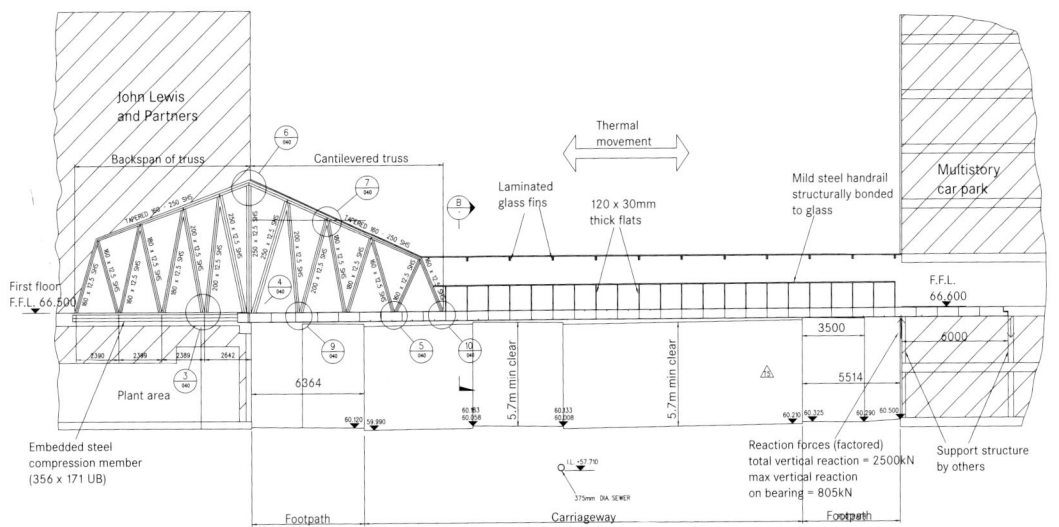

Elevation A-A

Labels within image:
John Lewis and Partners
6 040
Backspan of truss
Cantilevered truss
7 040
Thermal movement
Laminated glass fins
120 x 30mm thick flats
Mild steel handrail structurally bonded to glass
Multistory car park
B B
First floor F.F.L. 66.500
F.F.L. 66.600
4 040
3500
6000
9 040
5 040
10 040
5.7m min clear
5.7m min clear
5514
3 040
Plant area
6364
60.583 60.858
60.333 60.008
60.210 60.325
60.290 60.500
Embedded steel compression member (356 x 171 UB)
60.120 59.990
1L +57.710
375mm DIA. SEWER
Reaction forces (factored) total vertical reaction = 2500kN max vertical reaction on bearing = 805kN
Support structure by others
2390 | 2389 | 2389 | 2642
Footpath
Carriageway
Footpath

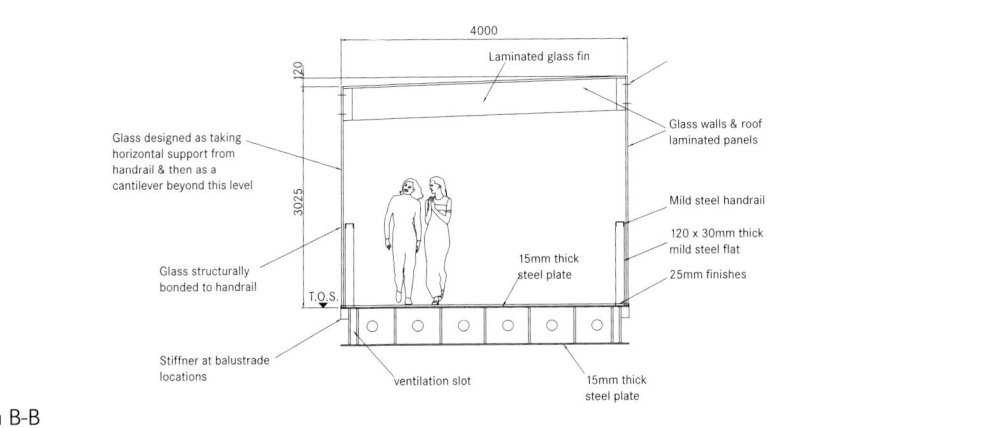

Section B-B

Labels within image:
4000
120
Laminated glass fin
Glass walls & roof laminated panels
Glass designed as taking horizontal support from handrail & then as a cantilever beyond this level
3025
Mild steel handrail
120 x 30mm thick mild steel flat
25mm finishes
Glass structurally bonded to handrail
15mm thick steel plate
T.Q.S.
Stiffner at balustrade locations
ventilation slot
15mm thick steel plate

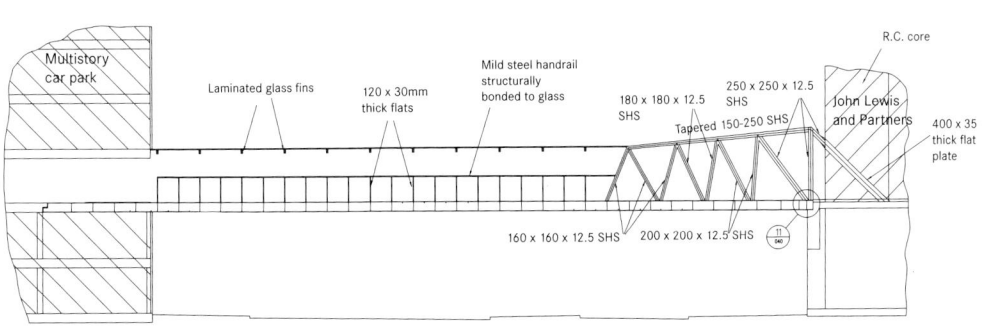

Elevation C-C

Labels within image:
Multistory car park
Laminated glass fins
120 x 30mm thick flats
Mild steel handrail structurally bonded to glass
180 x 180 x 12.5 SHS
250 x 250 x 12.5 SHS
R.C. core
Tapered 150-250 SHS
John Lewis and Partners
400 x 35 thick flat plate
160 x 160 x 12.5 SHS
200 x 200 x 12.5 SHS
11 040

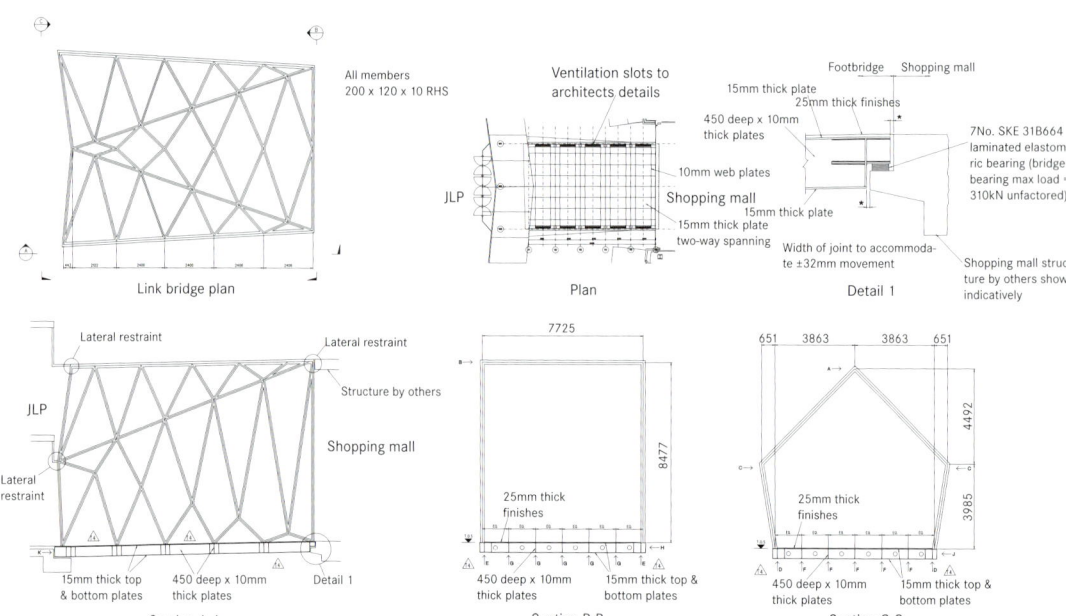

All members
200 x 120 x 10 RHS

Link bridge plan

Ventilation slots to
architects details

10mm web plates

JLP

Shopping mall

15mm thick plate
15mm thick plate
two-way spanning

Plan

450 deep x 10mm
thick plates

Footbridge   Shopping mall

15mm thick plate
25mm thick finishes

15mm thick plate

Width of joint to accommodate ±32mm movement

7No. SKE 31B664
laminated elastome-
ric bearing (bridge
bearing max load =
310kN unfactored)

Shopping mall struc-
ture by others shown
indicatively

Detail 1

Lateral restraint

Lateral restraint

JLP

Structure by others

Shopping mall

Lateral
restraint

15mm thick top
& bottom plates

450 deep x 10mm
thick plates

Detail 1

Section A-A

7725

8477

25mm thick
finishes

450 deep x 10mm
thick plates

15mm thick top &
bottom plates

Section B-B

651   3863   3863   651

4492

3985

25mm thick
finishes

450 deep x 10mm
thick plates

15mm thick top &
bottom plates

Section C-C

Footbridge section and details

Converting the Lionel Robbins Building into a library for the London School of Economics involved the refurbishment, alteration and extension of the former newspaper warehouse. Foster & Partners' scheme involved the creation of a new central atrium into which a central spiral ramp and lift structure were inserted to connect the floors. Additional work included partial demolition of the ground floor to create a double height void over the basement reading room, the provision of a new fifth floor, and the introduction of dense compact shelving which required strengthening the existing structure. AKT's commission involved undertaking investigations to establish the form and condition of the existing structure, and the development of a programme for concrete repairs. Their research uncovered high levels of carbonation in the concrete elements which had resulted in distress and a loss of strength in the floor structures, but found that repairs, rather than demolition, were possible. The project also includes a footbridge connection to an adjacent building; a glazed, monocoque structure that was entirely prefabricated off-site and lifted into place as a single piece.

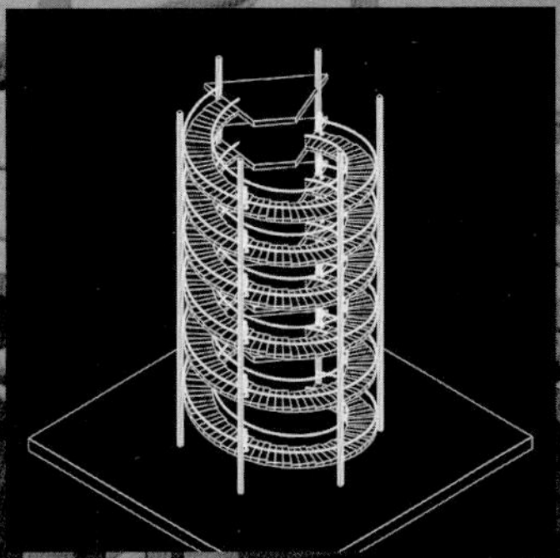

SPIRAL RAMP
LONDON SCHOOL OF ECONOMICS

FOSTER & PARTNERS
LONDON, 2001
CLIENT: LONDON SCHOOL OF ECONOMICS

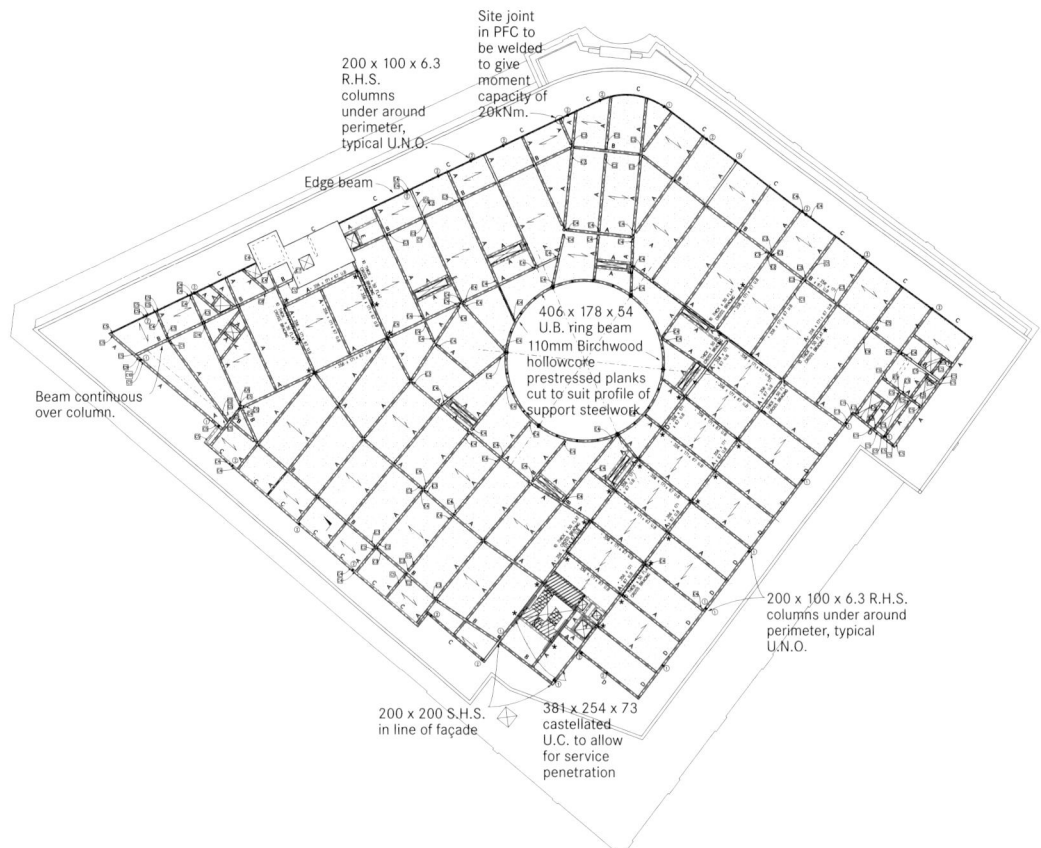

Plan of the building

This project had to confront a proposal that called for the demolition of an existing building that was infected with 'carbonated concrete', a common problem with concrete from the 1960s. This occurs when carbon dioxide penetrates into its pores and corrodes the steel reinforcement. While it is common to demolish buildings with such extensive defects, the Foster and Partners scheme relied on retaining the structure and making some significant interventions to improve the use of the building.

Would the concrete still stand up? Doesn't it become brittle?

We persuaded the client to spend part of the budget on material tests and research, in order to find out the strengths and weaknesses of carbonated concrete. This confirmed that the planned uses of the refurbished building could take advantage of some positive aspects of this decaying condition. At certain points of the decay, the concrete becomes strong in compression, but it also becomes more fragile. In response, we came up with a scenario where we didn't need to remove this structure completely, but to reinforce it where needed. Today, it is very difficult to develop ideas from independent research, and one has to use projects like this as far as is practical to learn. The result was that the budget for repairs could now be re-distributed to other more valuable interventions.

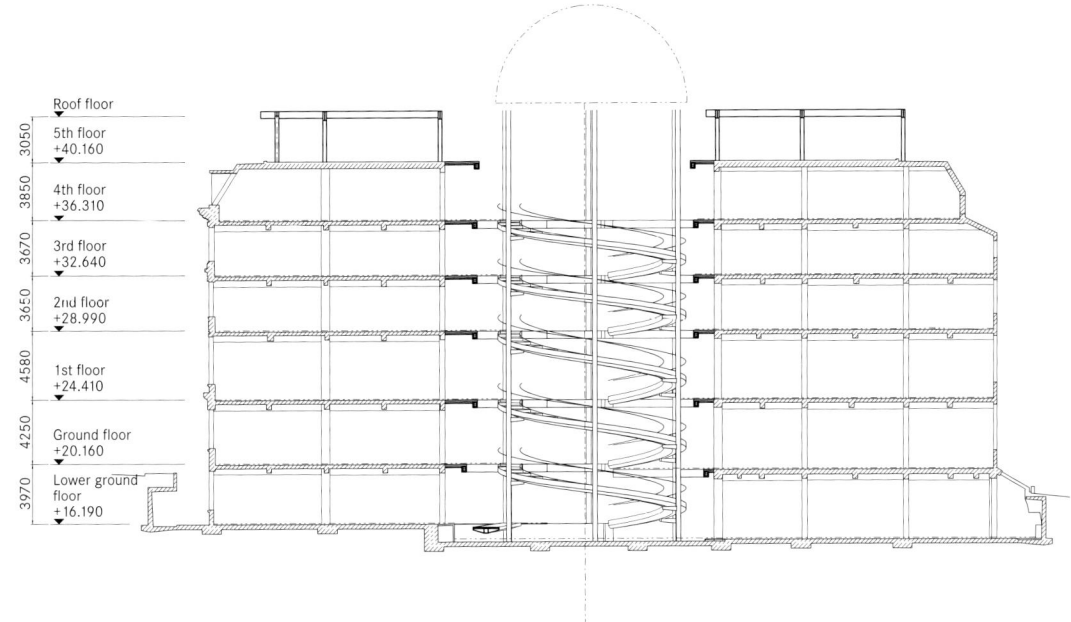

Roof floor
5th floor +40.160
4th floor +36.310
3rd floor +32.640
2nd floor +28.990
1st floor +24.410
Ground floor +20.160
Lower ground floor +16.190

3050
3850
3670
3650
4580
4250
3970

Section of the building

The strategy was to deliver a primary relational space in the form of a spiral ramp, situated in the light well at the centre of the building, which would also allow visitors to easily locate themselves within, and move through, the library. Fosters and Partners were working on this around the time of the Berlin Reichstag and New City Hall in London, where they have successfully used vertical circulation as a new programmatic device.

The spiral is made out of composite concrete with a steel plate that can be seen underside.

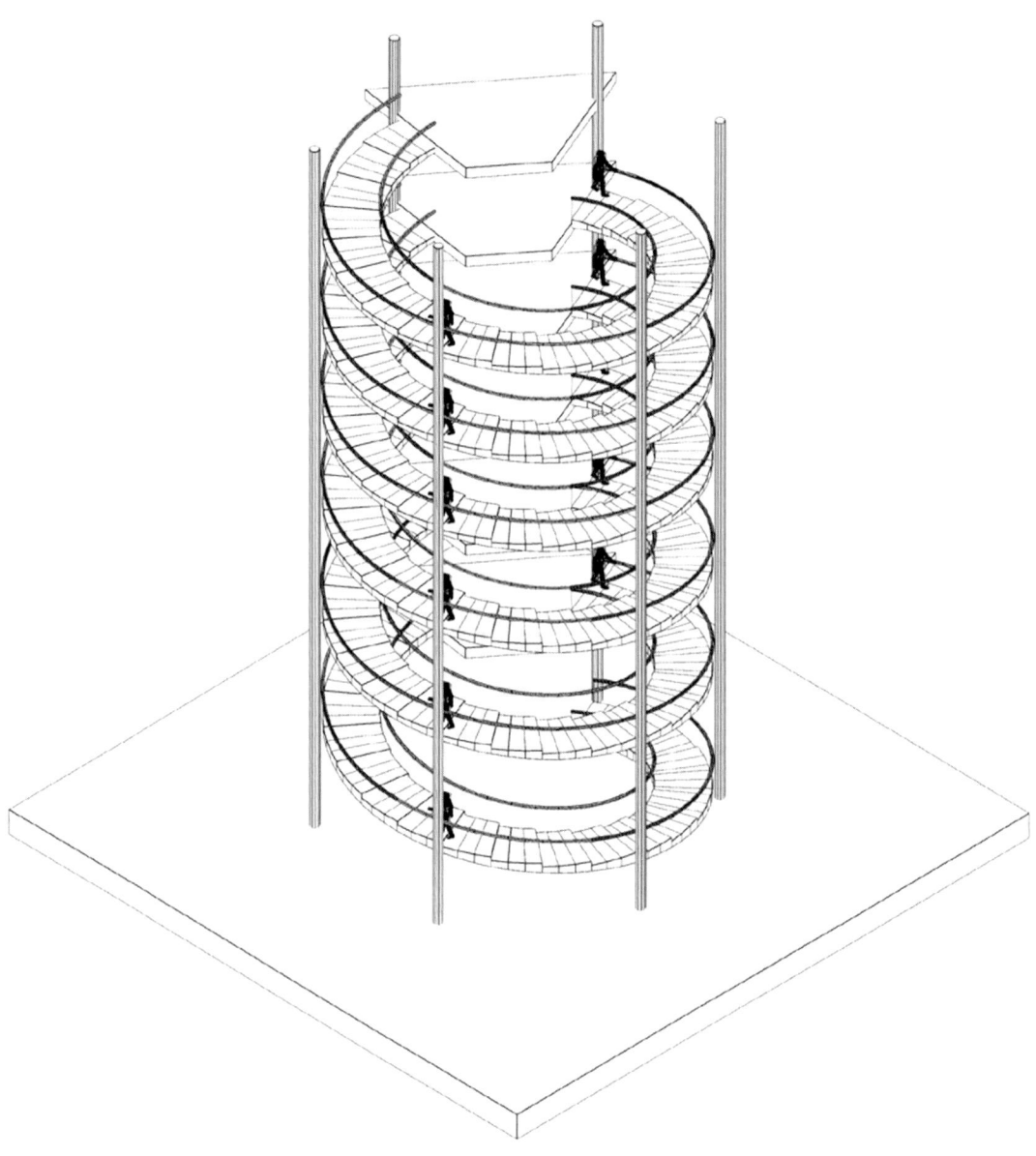

General arrangement

The geometry of the ramp provides stiffness and six new columns were inserted to support this independently, therein not departing any loads onto the existing structure. This self-stabilising system also allowed us to construct it with ease within the constraints of the existing building and site.

At the top, the light well terminates into a half dome skylight that directs natural light through the building. This dome was developed and made in the same way as the Peckham pods; in fact, by the same fabricator, Conley Construction.

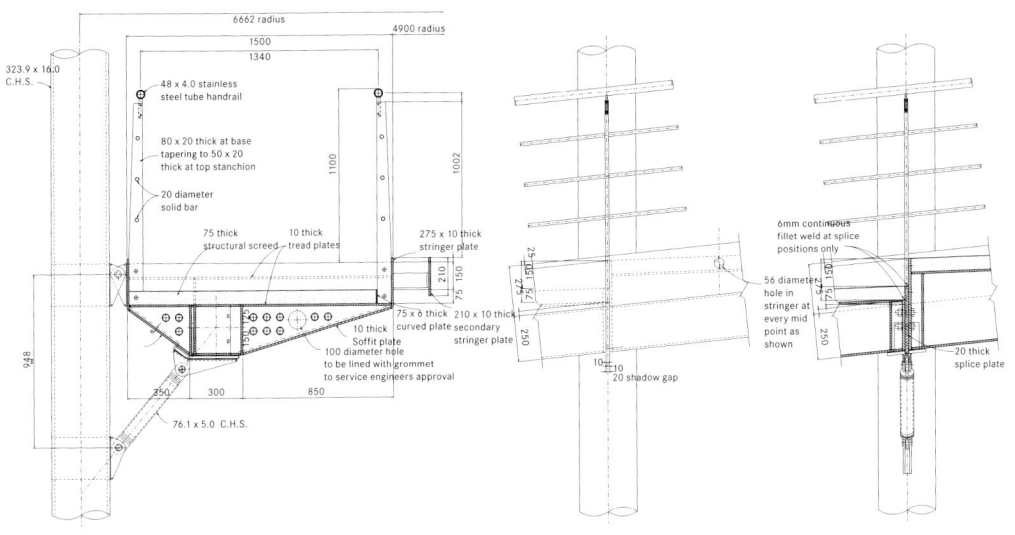

Section 1-1

6662 radius
4900 radius
1500
1340
323.9 x 16.0 C.H.S.
48 x 4.0 stainless steel tube handrail
80 x 20 thick at base tapering to 50 x 20 thick at top stanchion
20 diameter solid bar
75 thick structural screed
10 thick tread plates
1100
1002
275 x 10 thick stringer plate
210
75
150
10 thick curved plate
75 x 6 thick secondary stringer plate
210 x 10 thick Soffit plate
100 diameter hole to be lined with grommet to service engineers approval
948
350
300
850
76.1 x 5.0 C.H.S.

Section 2-2

25
151
52
275
250
10
10
20 shadow gap

Section 3-3

6mm continuous fillet weld at splice positions only
56 diameter hole in stringer at every mid point as shown
25
151
52
250
20 thick splice plate

View 1 from top

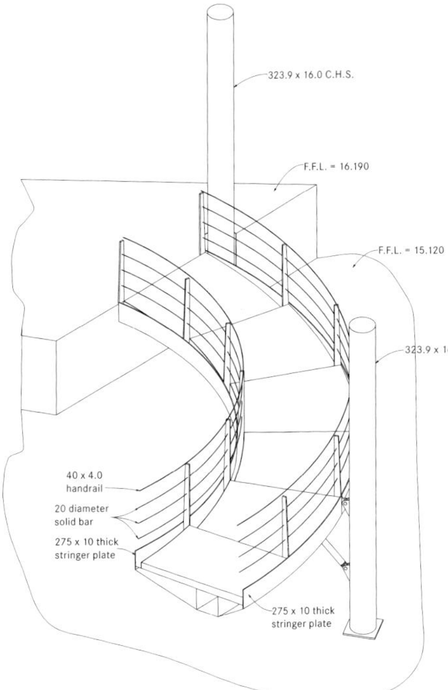

323.9 x 16.0 C.H.S.
F.F.L. = 16.190
F.F.L. = 15.120
323.9 x 16.0 C.H.S.
40 x 4.0 handrail
20 diameter solid bar
275 x 10 thick stringer plate
275 x 10 thick stringer plate

View 1 from underside

View 2

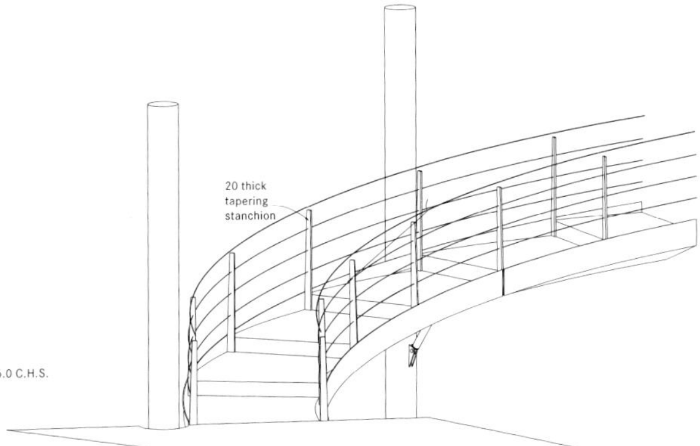

20 thick tapering stanchion

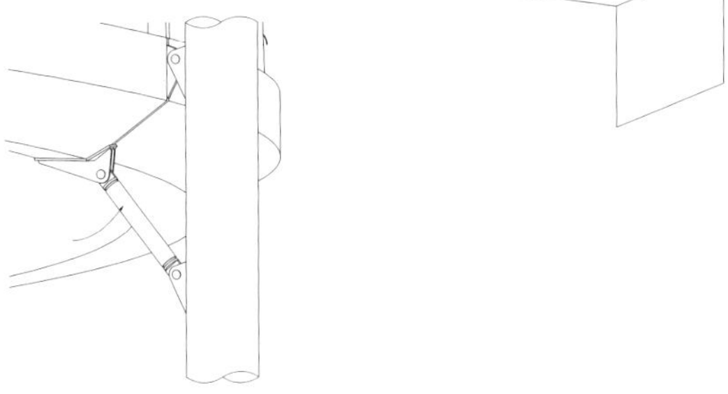

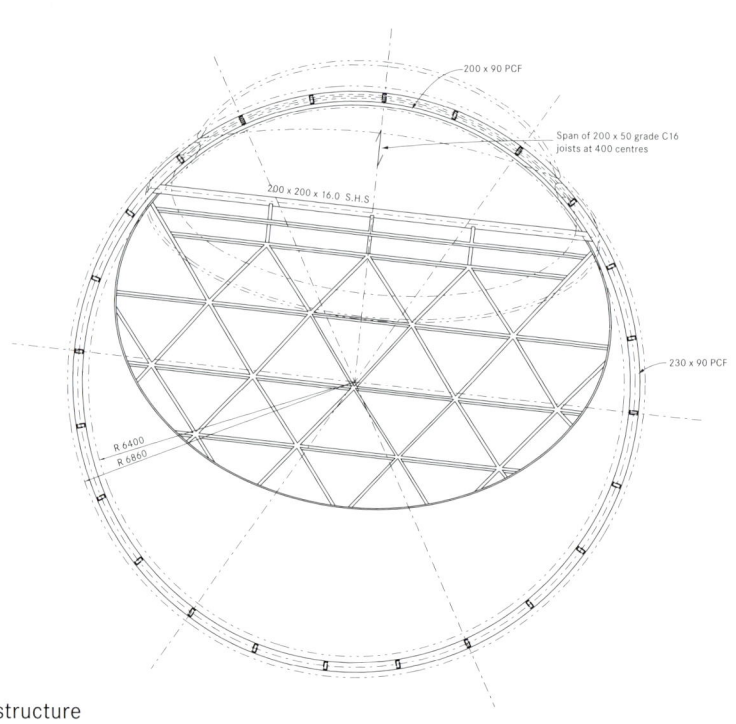

Plan of the dome structure

Labels visible on plan: 200 x 90 PCF; Span of 200 x 50 grade C16 joists at 400 centres; 200 x 200 x 16.0 S.H.S; 230 x 90 PCF; R 6400; R 6860

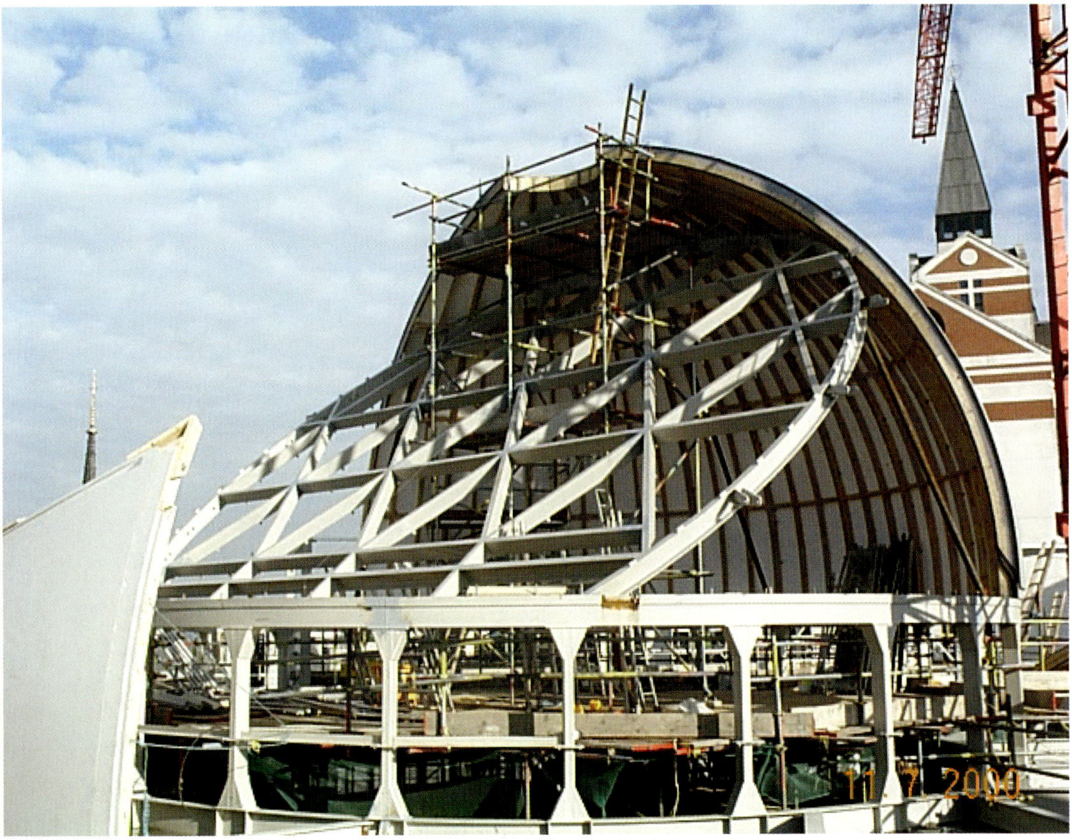
11 7 2000

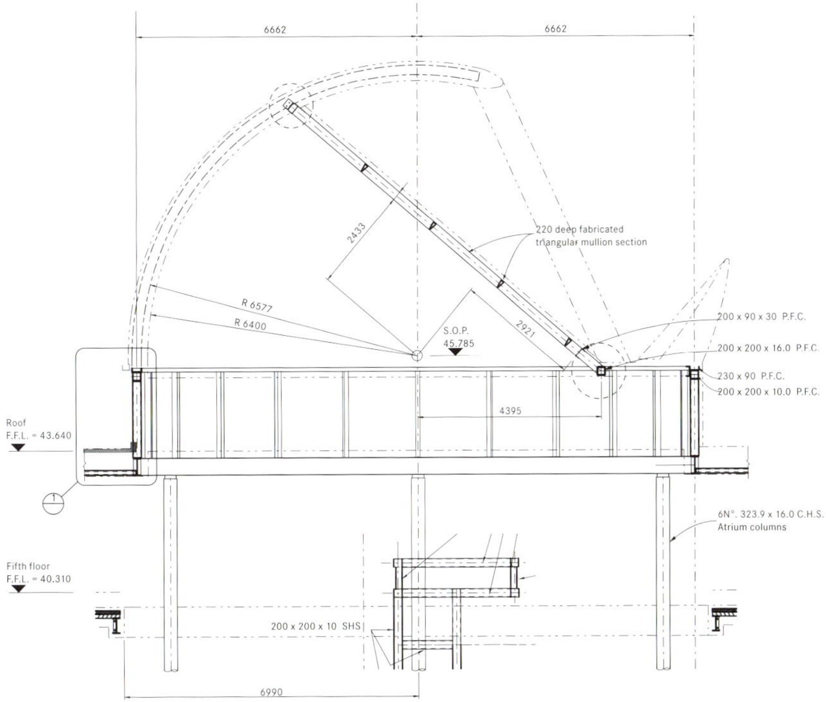
6662
6662
220 deep fabricated triangular mullion section
2433
R 6577
R 6400
S.O.P. 45.785
2921
200 x 90 x 30 P.F.C.
200 x 200 x 16.0 P.F.C.
230 x 90 P.F.C.
200 x 200 x 10.0 P.F.C.
4395
Roof F.F.L. = 43.640
1
6N°. 323.9 x 16.0 C.H.S. Atrium columns
Fifth floor F.F.L. = 40.310
200 x 200 x 10 SHS
6990

Section of the dome

The other thing that we like to highlight about the spiral diagram is that it metaphorically captures how our office operates. When we were working on this project we started to be a bit literal about the importance of natural forms and natural ideas in combination with organisational methods in the office. Our first office photograph was in fact taken on this spiral.

AKT office members photographed in the spiral

The redevelopment of Walbrook Square, a 3.7-acre site in the City of London, stems from a partnership between the architectural practices of Jean Nouvel and Norman Foster, with AKT as structural engineers. The diverse context contains the remains of the Roman Temple of Mithras and is a present-day transportation nexus. Composed of four buildings, the project includes approximately 300,000 square metres of retail, restaurant, and office space. The complex comes together to form a new public space and streets that are based on historic routes. A rational, nine by twelve-metre structural grid, above a subterranean concrete base, gives maximum flexibility to the office spaces. Two towers are capped with 'cloud-like' forms AKT used 3-D digital modelling to examine exactly how the shapes relate to each other and subsequently designed a responsive structural system.

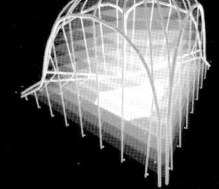

CLOUDS, ROOF STRUCTURE
WALBROOK SQUARE

ATELIER FOSTER NOUVEL
LONDON, 2006-
CLIENT: STANHOPE PLC

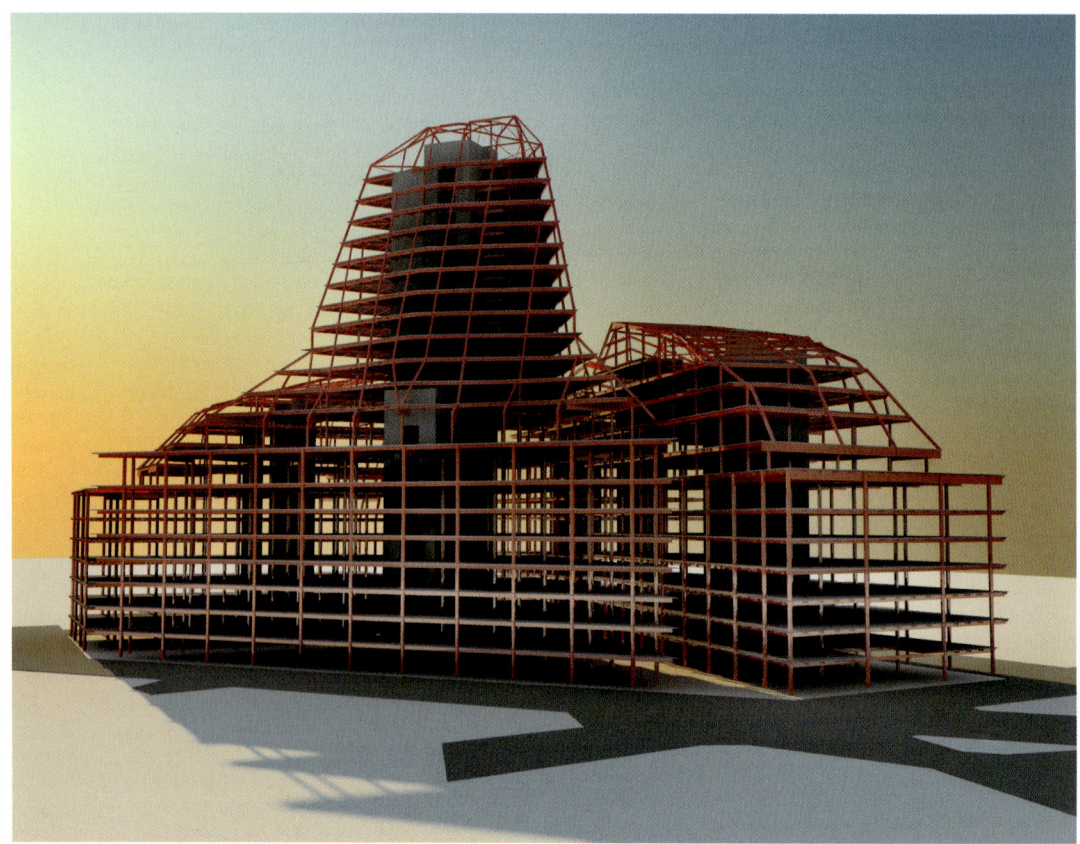

Frame model interactively used to iterate

The title of this book, Design Engineering, was based on the idea of defining some kind of new discipline. Is the aesthetic section of it partly being able to inhabit the minds of the architects that you are working with, or understand what they are trying to achieve through their eyes?

By collaborating with architects you share that pain with them. And you can't say you can do that for everyone; you have to understand that person's derivative position first.

It's a form of empathy?

No one will work with engineers who are not technically competent. But if that is all a firm can offer, without any aesthetic empathy, or anything else, then what makes them stand out? All of the staff in our office is recruited with this mentality.

Assembly in existing building

Models for planning consultation. Atelier Foster Nouvel

Building 1
S.S.L. 17.050

Building 2
S.S.L. 16.450

Building 4
S.S.L. 17.050

Building 3
S.S.L. 16.850

Massing and topology

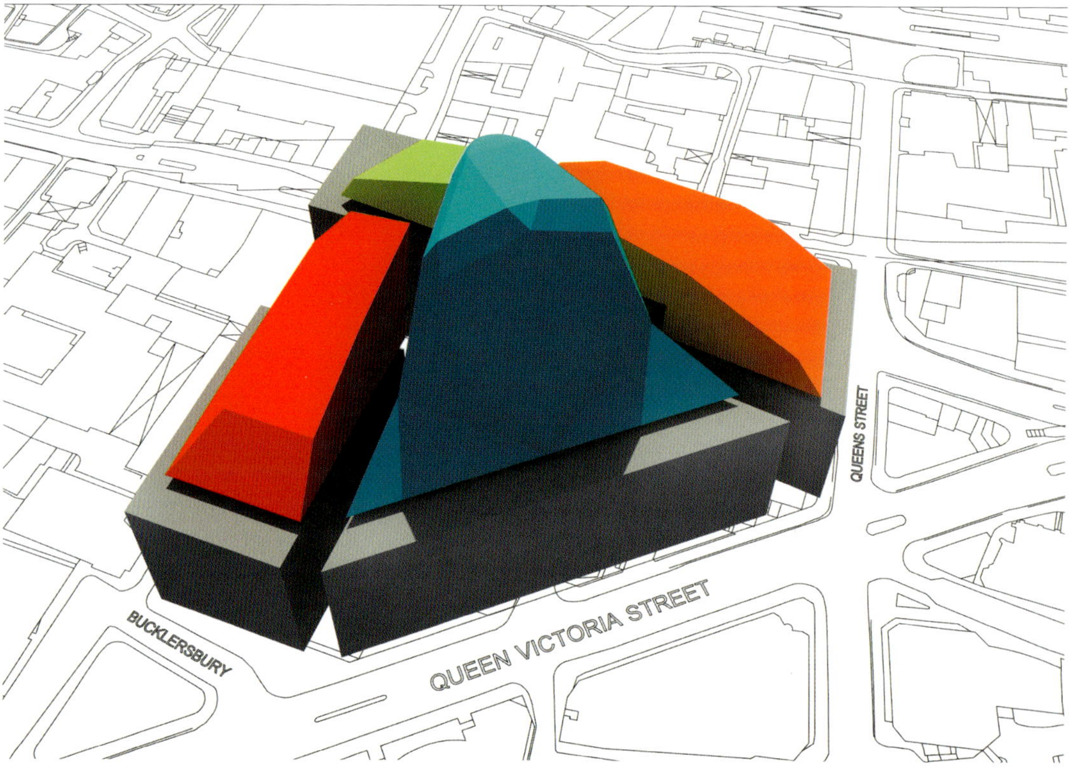

QUEENS STREET

BUCKLERSBURY

QUEEN VICTORIA STREET

Public realm strategy for form finding. Atelier Foster Nouvel

On Walbrook Square you worked with both Foster & Partners and Ateliers Jean Nouvel. How was this dual collaboration?

It is fascinating and we think that many such collaborations will be formed in the future, as architects share knowledge to resolve some of the most complex architectural challenges.

Walbrook Square is the major project in the heart of London right now, the last of the large island sites. One aspiration for the site is to develop it into the most attractive commercial real estate in the world. Conceptually, the whole idea examines how the city develops new buildings. We are fighting to have floor space but also fighting away from the 'tower typology' in this project, in order to explore typologies that are close to the ground, rather than fly in the air.

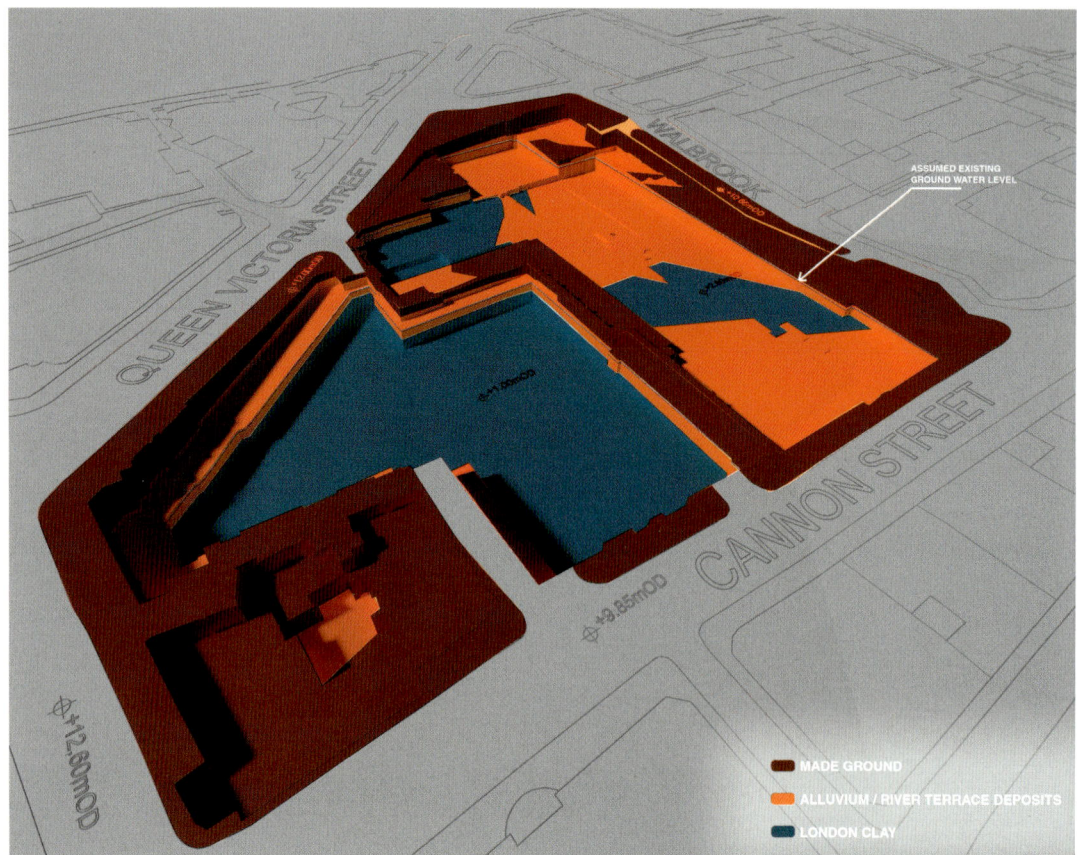

The site is constrained by ground conditions, surrounding buildings, St. Paul's protected views and a number of existing underground lines. The shapes are really driven by protected viewing corridors, leaving a developable envelope which this building takes up, before it changes at about the eleventh floor. It responds to the street at the lower level and then at the higher level to the sky, with varying heights when viewed from the different streets.

Though the cloud forms draw a lot of attention, there is a real challenge below ground, because this site has been piled twice in the last century. The forms therefore are also constrained by where we can fit new piles (a problem that is going to reoccur in many West European cities where the sites have been developed several times in the last hundred years). We've inherited this problem from the 1960s and there is little space to fit new foundations. The older buildings and construction methods were such that smaller structural grids and a large number of small diameter piles were the answer. The advent of long span steel construction means less columns but often larger diameter piles that have to avoid existing pile foundation. So this 'bottom-up' challenge has been key to the form-making process for the architects.

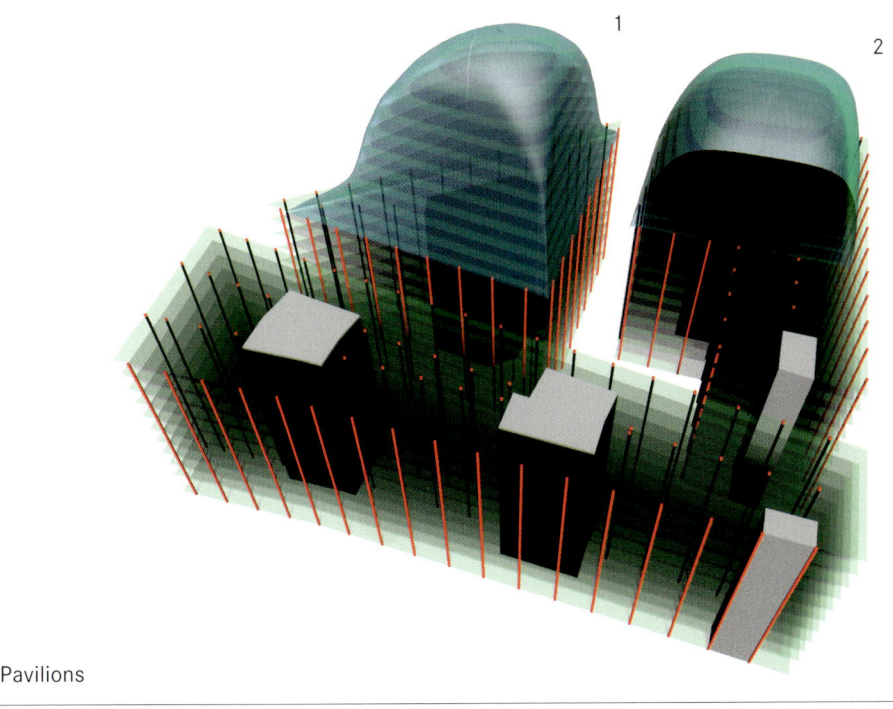

Pavilions

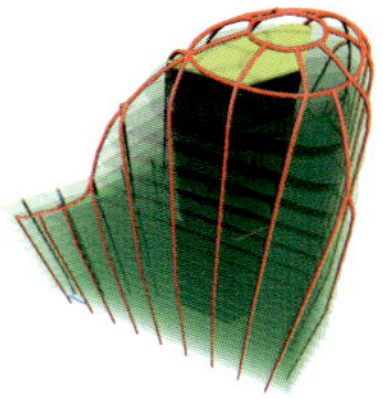

1: v2-1

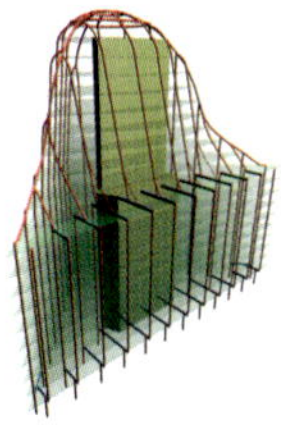

1: v2-2

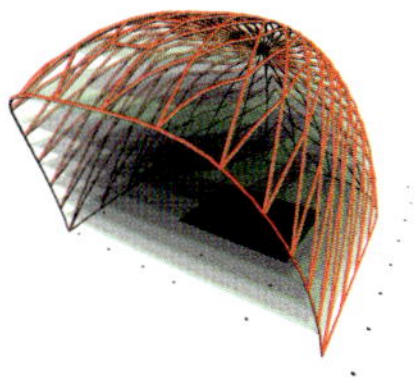

2: Option

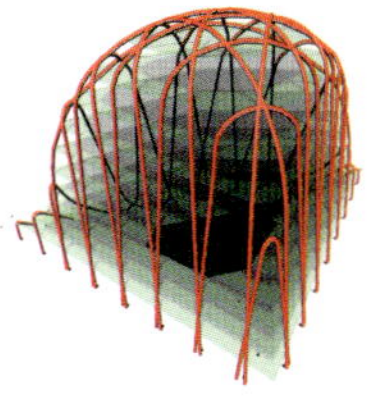

1: Option 1-1

1: Option 3-1

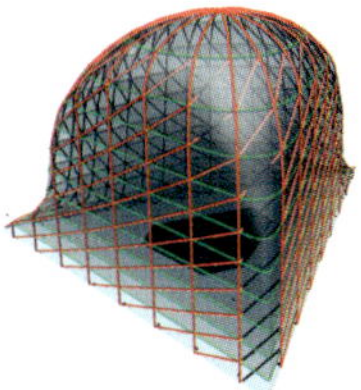

1: Option 2-1

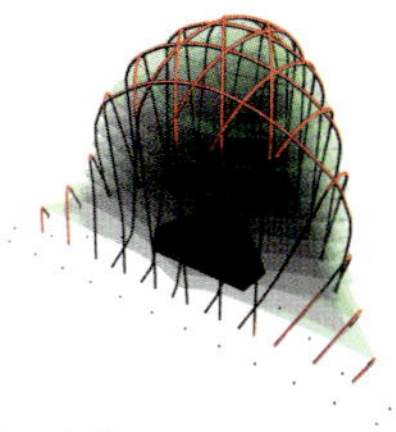

1: Option 1-2

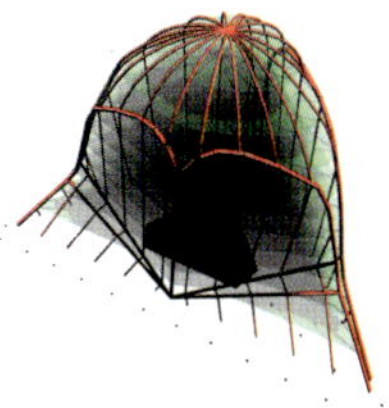

1: Option 3-2

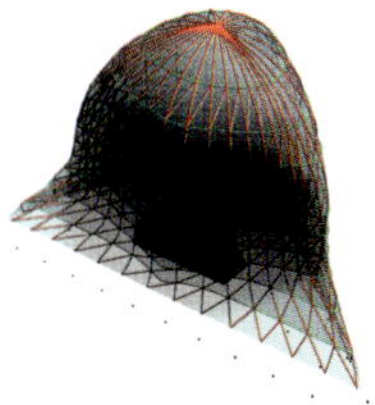

1: Option 2-2

Option studies for frames

Models for site. Atelier Foster Nouvel

The architects have formed a project office under one roof to focus on this building so that the project is developed jointly by both offices. For us, this collaboration of two architects is exciting and presents many new potentials. Our work is scrutinised at every level, and to get to a really exemplar solution we have to work in three dimensions at speed.

The representations, analysis and synthesis of ideas takes another dimension, driven by the pace and scale. The upper parts of all the structures are transparent and the structural frame has been carefully coordinated to respond to the 'chiselled' forms whilst co-ordinating carefully with the façade patterns.

Project visualisation, Atelier Foster Nouvel

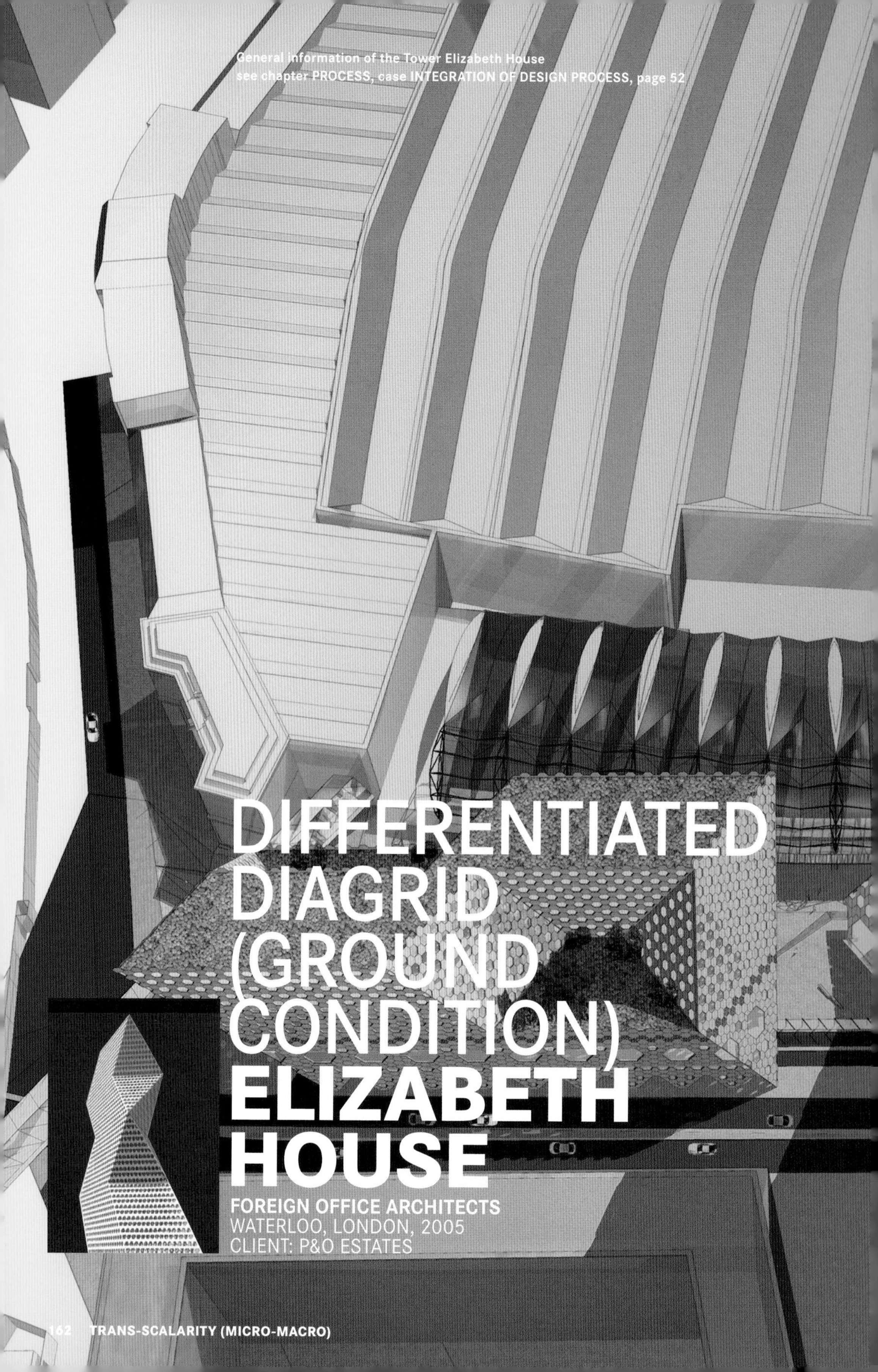

General information of the Tower Elizabeth House
see chapter PROCESS, case INTEGRATION OF DESIGN PROCESS, page 52

DIFFERENTIATED DIAGRID (GROUND CONDITION)

ELIZABETH HOUSE

FOREIGN OFFICE ARCHITECTS
WATERLOO, LONDON, 2005
CLIENT: P&O ESTATES

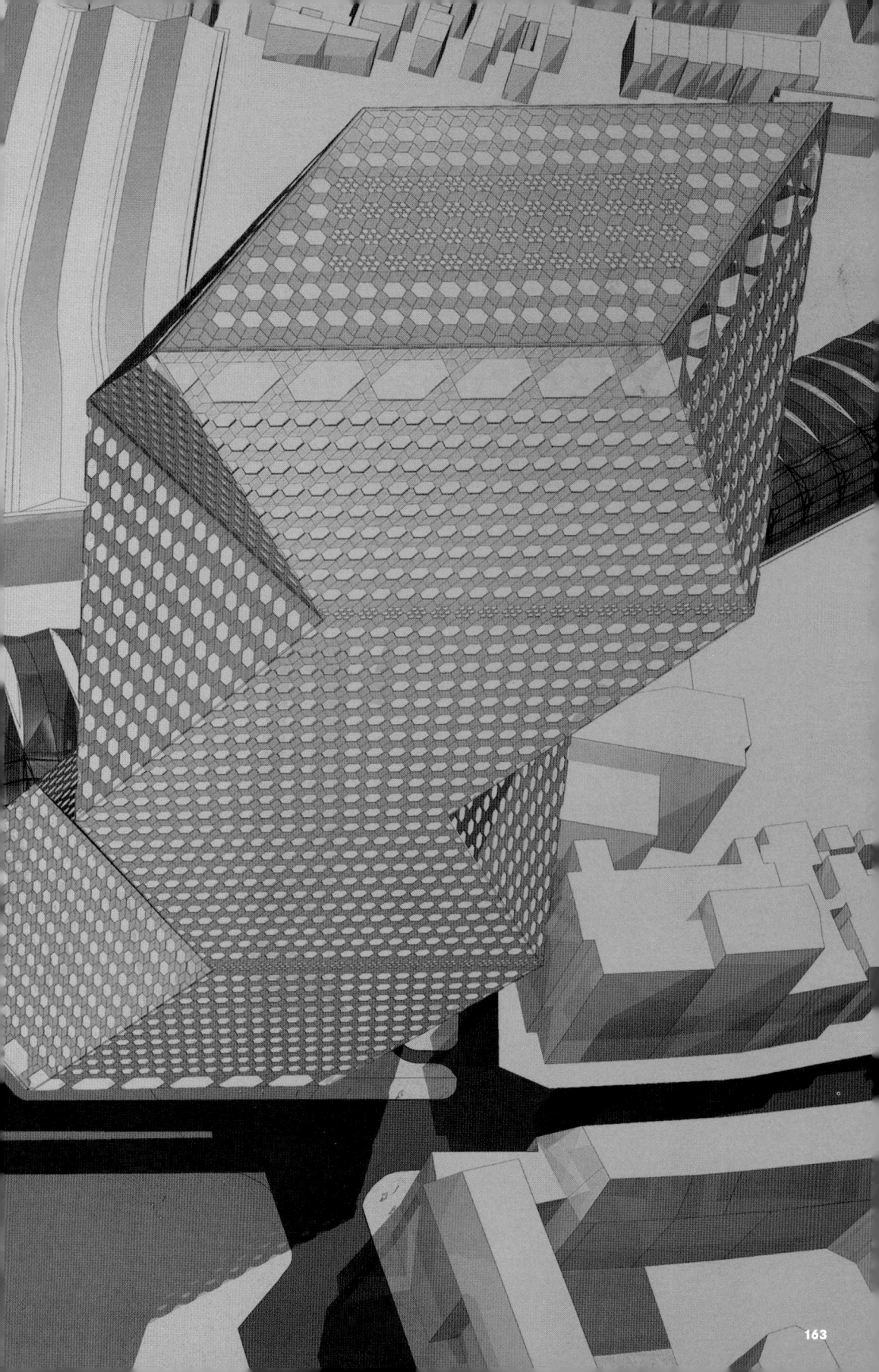

You talked about tall buildings and recent competitions that have been interesting, in particular the competition for Elizabeth House. What was the driving force for the structural concept on this?

This central London site is at a 'transport hub' in a very congested and constrained site directly to the north of Waterloo Station. The site has an existing building that is to be dismantled, and the client P&O wanted to redevelop a tall building on it. The arrangement of underground tunnels and their condition play a key role in making a new project here. The tunnels traverse the site at various levels under the site, and had to remain undisturbed throughout the works.

So this meant that the magnitude and location of the vertical loads had to be carefully organised.

The protected views and site constraints led FOA to propose a trapezoidal plan, with sixteen-floor modules that tessellate to respond to views from various parts of London, but also to land on the narrow footprint. This allowed us to propose a diagrid column arrangement on the two main elevations. This grid was selected to support conventional composite floor beams that span on to it from the core. Simultaneously the diagrid provides an ideal primary system for the hexagonal opaque panels of the façade.

How did the diagrid become differentiated in some of the models?

Competitions often allow some room to demonstrate ways of thinking, so a number of models were made. One included a grid that becomes increasingly perforated as one gets higher up the building where vertical loads are reduced. In the final entry FOA were not so keen on the differentiation but it allowed us as a team to show the process, even considering some irregular differentiation.

The final proposal was very efficient in that the diagrid not only transfers loads vertically but was also able to contribute to the lateral stability of the building by working with the floor, as well as providing transfer girders close to the ground where needed.

FOA's visualisation

EXTREMES

The capacity to work with a large number of extremely varied projects in scale, types and architectural positions puts a new demand on the engineering response. It is an inevitable position that we encountered, and so have developed our specific approach to cope with this position. Such innovations in approach or solutions become part of the environment.

For instance, sometimes advances can result from the complexities of large projects. Alternatively, small projects can also become tools for investigation that eventually inform other projects at a much larger scale. In fact, sometimes the most fruitful innovations develop from the smallest projects. It is this dichotomy between the simple and the complex, whereby the large and the small can positively affect one another, that produces a ground for innovations across projects.

Other extremes such as working in a variety of international projects where skills and processes vary also allow us to develop skills and habits that are not defined easily in mainstream structural engineering. This is a particular condition of contemporary practice where new and young practitioners are today able to work on specific projects across the world due to their agility and accessibility. Similarly, at the 'micro' and 'macro' scales as well as with complex and simple programmes, demonstrates a clear value and power of working with extremes. Extremes in technologies of form and process are essential in enabling the right environment for innovation.

This café is designer/artist Thomas Heatherwick's first sortie into buildings after a host of structures and sculptures. Located on the beach at Littlehampton, a resort on England's south coast, it appears like a wave frozen in Corten steel. Forty meters long and six meters high, the structure is formed out of steel plates welded in strips to make a continuous surface. Although its shape is intuitively derived, AKT used finite element analysis to understand how it would perform. The curved steel acts as an arch, but AKT's analysis showed that the flattened area required extra ribbing, which was mounted perpendicularly to the main steel strips. Because sand is a good bearing material, and not prone to shifting and drifting on this site, simple concrete strip foundations were adequate to support the café.

STEEL STRUCTURE LITTLE-HAMPTON CAFÉ

THOMAS HEATHERWICK STUDIO
LITTLEHAMPTON, ENGLAND, 2007
CLIENT: BROWNFIELD CATERING

This project gave us an opportunity to push forward our knowledge of the 'low order' craft of steel plate construction and 'higher order' technologies in digital modelling software. When we discuss our work process we often deliberately place the Phæno Centre in Wolfsburg and the Littlehampton project next to each other, as together they draw out extremes of thinking, material, scale, budget site and architectural positions.

Was it a different process for you as an engineer to work with an artist?

We've often worked with artists, but this situation differed from the usual practice of taking art into architecture; we were instead taking architecture into what some define as sculpture or art. This was an unusual process for us as engineers because our practice is too controlled and optimal to deal with this approach to architecture, and so we learned a lot through this new type of engagement. It would be fair to say that the engineers' role is perhaps even wider but certainly more focused as the 'object' is a given and there is little opportunity to reshape the solution. Here our role is closer to simply finding the best way to analyse, rationalise and construct the building, to protect the initial metaphor which in this case was about an 'object of the sea', a piece of driftwood that has floated to the shore. In this role we were much more the enablers, assisting in setting the steps.

The metaphor of driftwood was the key concept

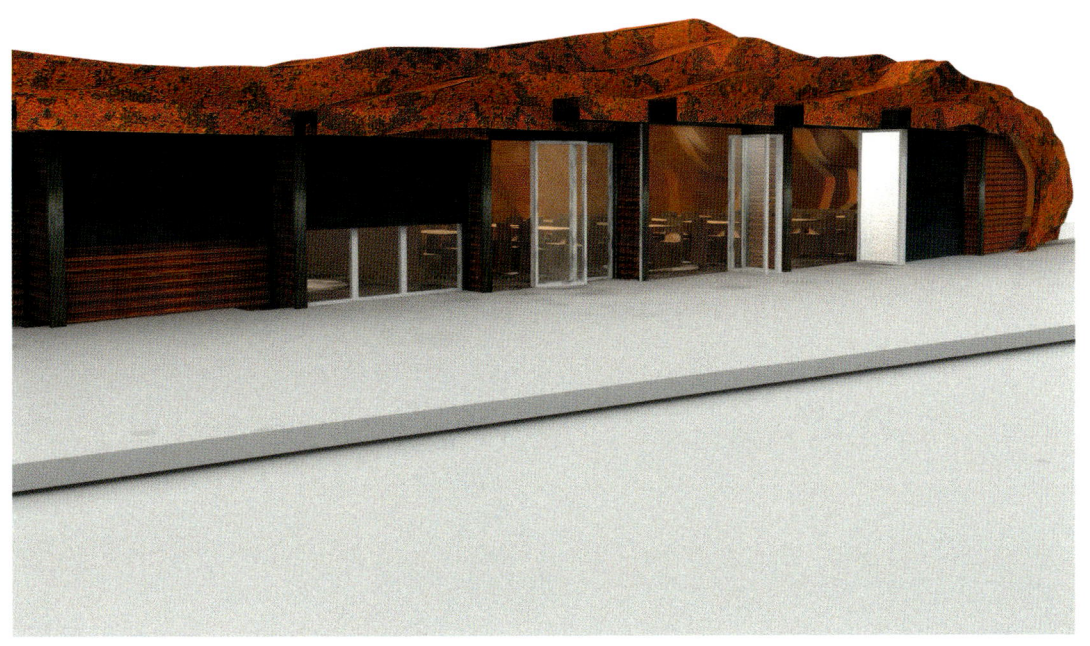

Seafront visualisation, open view to the sea

Shop fitting by Littlehampton Welding

The 'steel plates' were unfolded to find 'tiles' that could be welded together to form the raw exoskeletal structure, and we used 'digital projects' for the first time as a tool to use the same model for analysis of stresses and for the production of fabrication drawings. Littlehampton Welding were keen to assemble the structure in the fabrication shop to ensure that the splice points of the form would fit when they arrived on site. This type of prototyping and rigour ensured that site joints matched.

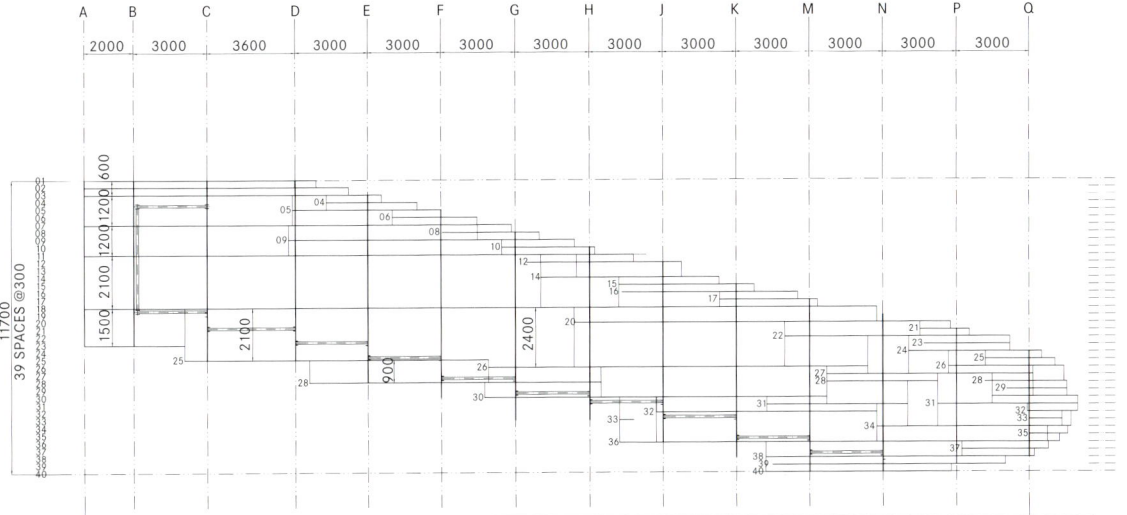

Sectional setting out of each 'ribbon'

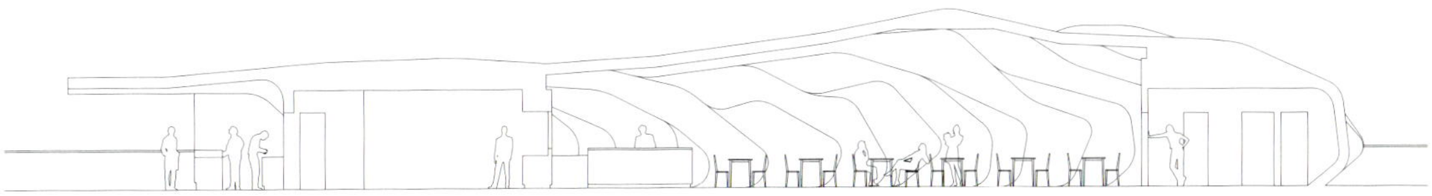

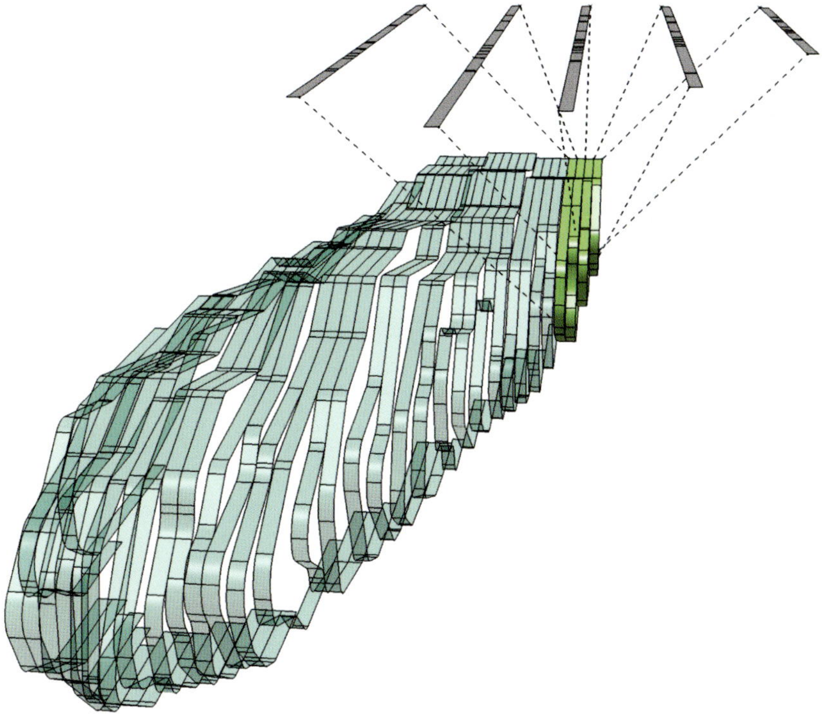

Catia model to rationalise the ribbons in order to determine the geometry for fabrication

Eric Parry's inclined obelisk is located at the southern end of London Bridge and the monument marks entry into the Southwark area, one of the city's oldest suburbs. Sixteen metres long and crafted out of twenty-five pieces of Portland stone, it stands at an angle of 18.5 degrees off-vertical, pointing towards Southwark Cathedral. To achieve the dramatic tilt, AKT employed a post-tension solution, rather than introducing reinforcement bars, which ensured the correct amount of friction between the stone pieces. Six post-tension rods of varying length reach almost to the top (the apex piece is glued on) and are embedded in a reinforced concrete base. The obelisk is part of an effort to regenerate northern Southwark, and the scheme includes a visitors centre comprised of a lightweight steel and timber structure that clips onto an existing concrete walkway.

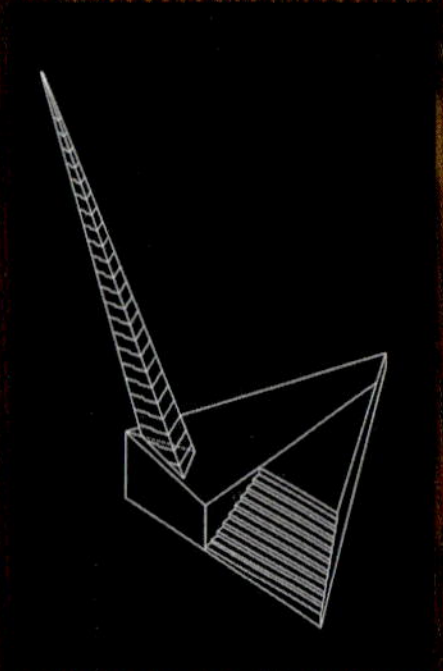

STONE POST-TENSIONING
SOUTHWARK GATEWAY

ERIC PARRY ARCHITECTS
SOUTHWARK, LONDON, 1999
CLIENT: LONDON BOROUGH OF SOUTHWARK

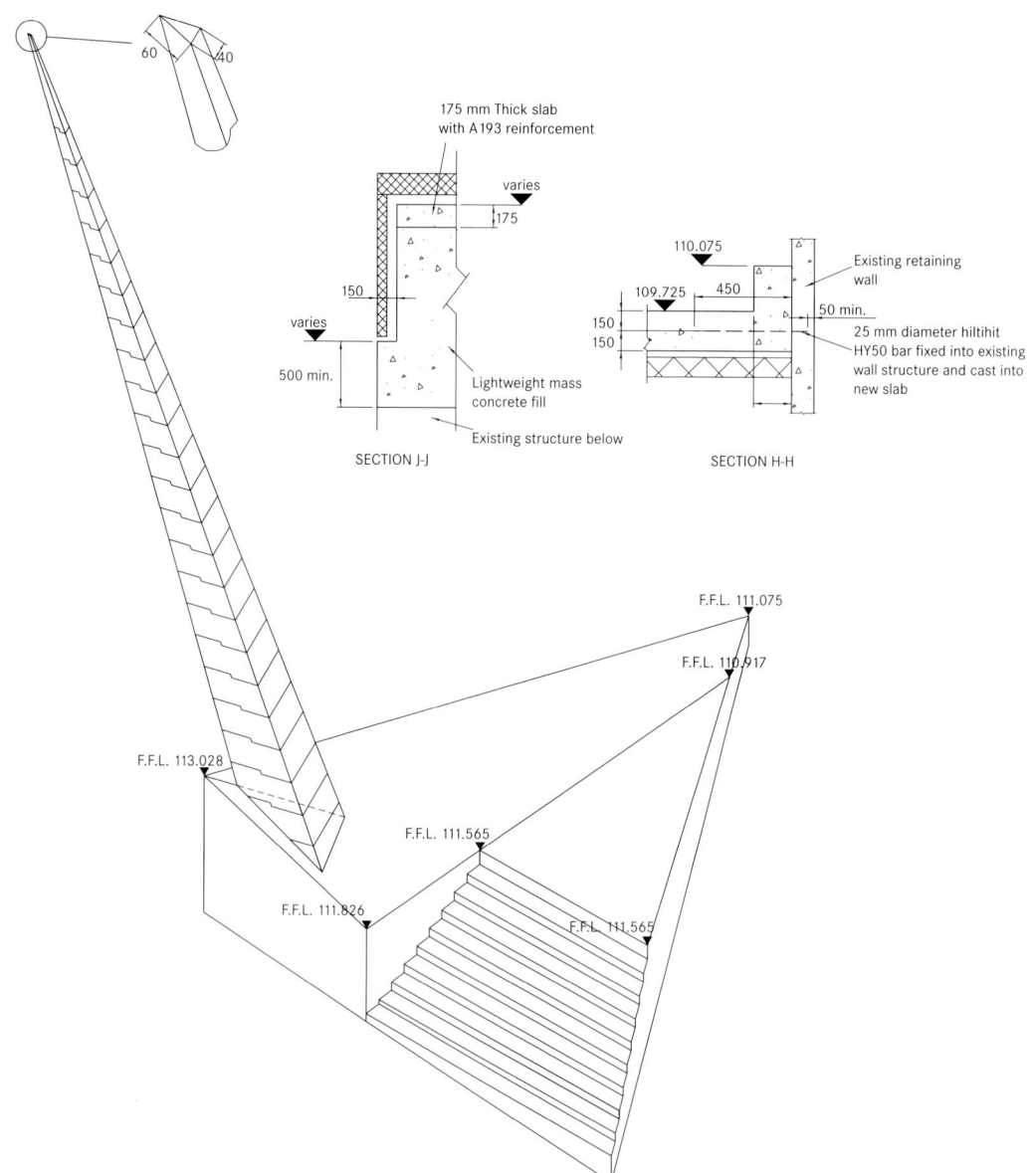

60   40

175 mm Thick slab
with A193 reinforcement

varies

175

150

varies

500 min.

Lightweight mass
concrete fill

Existing structure below

SECTION J-J

110.075

109.725   450

150
150

Existing retaining
wall

50 min.

25 mm diameter hiltihit
HY50 bar fixed into existing
wall structure and cast into
new slab

SECTION H-H

F.F.L. 111.075

F.F.L. 110.917

F.F.L. 113.028

F.F.L. 111.565

F.F.L. 111.826

F.F.L. 111.565

Isonometric view of slab levels to plinth

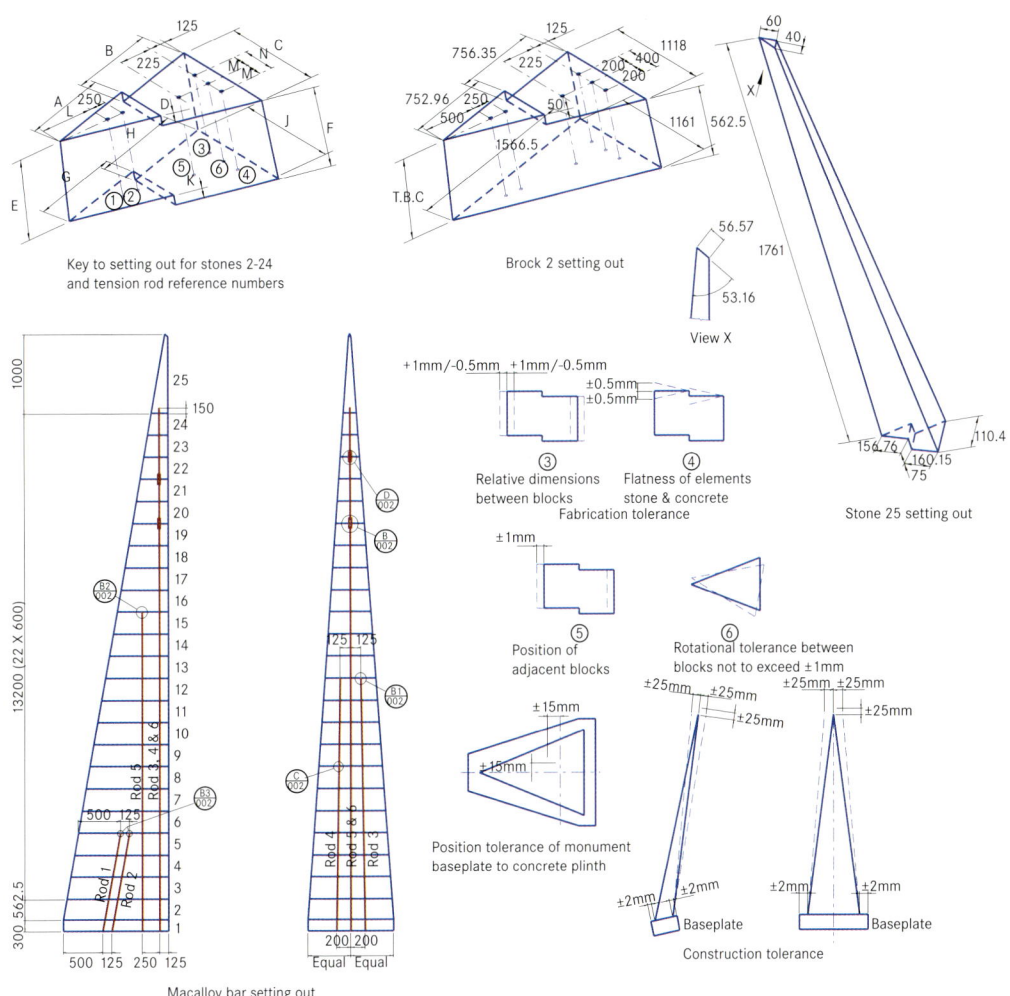

Key to setting out for stones 2-24
and tension rod reference numbers

Brock 2 setting out

View X

Stone 25 setting out

+1mm/-0.5mm +1mm/-0.5mm
±0.5mm
±0.5mm

③ Relative dimensions between blocks
④ Flatness of elements stone & concrete
Fabrication tolerance

±1mm

⑤ Position of adjacent blocks

⑥ Rotational tolerance between blocks not to exceed ±1mm

±15mm

±15mm

Position tolerance of monument baseplate to concrete plinth

±25mm ±25mm
±25mm
±2mm ±2mm
Baseplate

±25mm ±25mm
±25mm
±2mm ±2mm
Baseplate

Construction tolerance

Macalloy bar setting out

The only tangible brief that the architect had was to create a landmark for the borough of Southwark, which was going through a process of regeneration. The obelisk's tip had to be at a specific point in space for visual considerations, but the vertical projection of this point on the ground fell outside the given site boundaries. Therefore, the obelisk had to tilt from its springing point to reach this point in space.

The architect had a strong desire to explore the use of natural stone instead of the easier alternative of reinforced concrete. We decided to use large blocks of Portland stone that stack against each other, but in order to canter the tilt the whole form is positioned to the foundation. 'Keys' in each block provide shear resistance. What's interesting about this project is the fact we used Portland stone in a way that many people find difficult to understand, as the post-tensioned cables are not visible but located instead at the centre of the blocks. As the blocks taper to the top we had to terminate the post-tensioned cables at the penultimate piece and glue the last section.

Lucien Simon's sculpture is a symbol of cultural and urban renewal located on the southern side of Tower Bridge. The CIT Group developers commissioned the five metre-high crystalline artwork and AKT provided advice on structural configuration, element sizing, and material specification. The Teflon-coated glass sheets are held together by a steel rod running through the centre of the stack. It passes through a hole in each sheet of glass, which is large enough to allow moderate movement, without compromising stability.

GLASS POST-TENSIONING
LONDON BRIDGE SCULPTURE

KSS ARCHITECTS
LUCIEN SIMON, artist
SOUTHWARK, LONDON, 1999
CLIENT: THE CIT GROUP

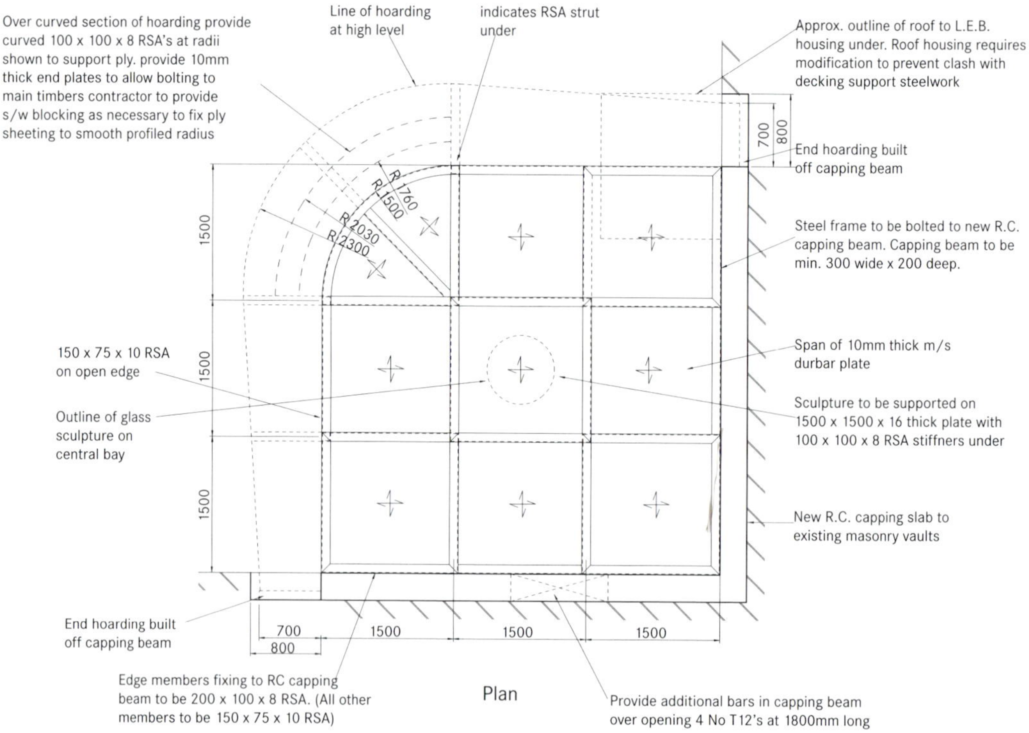

Over curved section of hoarding provide curved 100 x 100 x 8 RSA's at radii shown to support ply. provide 10mm thick end plates to allow bolting to main timbers contractor to provide s/w blocking as necessary to fix ply sheeting to smooth profiled radius

Line of hoarding at high level

indicates RSA strut under

Approx. outline of roof to L.E.B. housing under. Roof housing requires modification to prevent clash with decking support steelwork

End hoarding built off capping beam

Steel frame to be bolted to new R.C. capping beam. Capping beam to be min. 300 wide x 200 deep.

R.1760
R.1500
R.2030
R.2300

1500

150 x 75 x 10 RSA on open edge

1500

Outline of glass sculpture on central bay

1500

Span of 10mm thick m/s durbar plate

Sculpture to be supported on 1500 x 1500 x 16 thick plate with 100 x 100 x 8 RSA stiffners under

New R.C. capping slab to existing masonry vaults

End hoarding built off capping beam

700
800

700   1500   1500   1500
800

Plan

Edge members fixing to RC capping beam to be 200 x 100 x 8 RSA. (All other members to be 150 x 75 x 10 RSA)

Provide additional bars in capping beam over opening 4 No T12's at 1800mm long

Lucien Simon works a lot with glass at a smaller scale, and when he came to us with this proposal we were motivated by the idea of making a glass spike that, as a piece of sculpture, didn't need to comply with any specific codes. In this situation, we couldn't actually conduct a full analysis and to some extent relied on imitation to the point that, in fact, the prototype was the final object. The post-tensioning technique that we had already used in the Southwark Gateway needle is used again here. Both projects present a very precise and particular kind of extreme problem in the way that steel works with either stone or glass, without putting too much stress on the material. The glass sheets are glued together and then post-tensioned by a central tie rod.

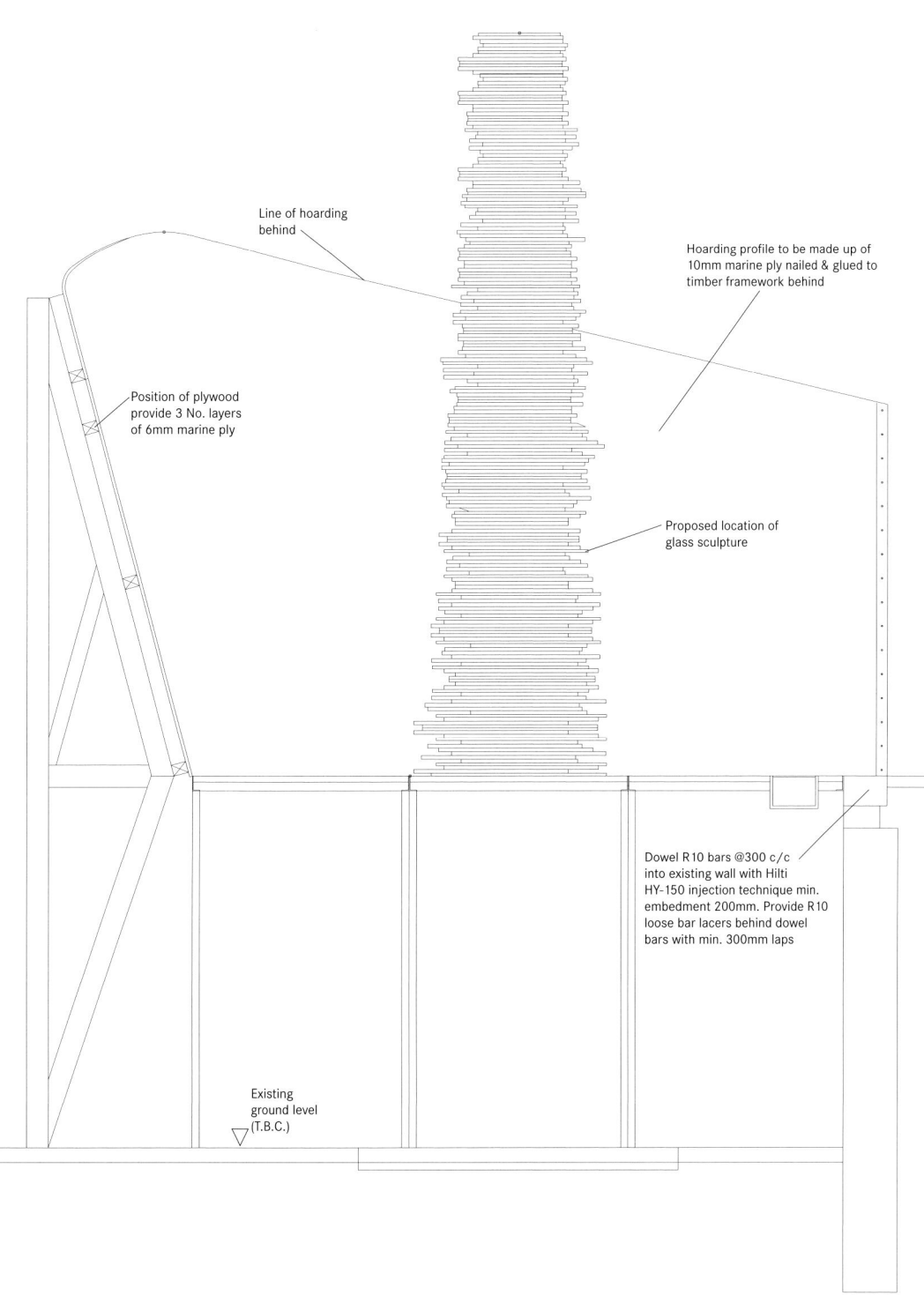

Line of hoarding behind

Hoarding profile to be made up of 10mm marine ply nailed & glued to timber framework behind

Position of plywood provide 3 No. layers of 6mm marine ply

Proposed location of glass sculpture

Dowel R10 bars @300 c/c into existing wall with Hilti HY-150 injection technique min. embedment 200mm. Provide R10 loose bar lacers behind dowel bars with min. 300mm laps

Existing ground level (T.B.C.)

For this winning competition entry designed by FOA, AKT proposed using a steel truss system to shape the folded surface that defines the building's exterior and spans each of the two concert halls. A second steel frame, that services the auditoria, fits within the primary structural system and supports the large spans of glass at the end of each hall. The volumes sit on a concrete podium and appear to "float" above the ground plane.

WRAPPING FRAME AND FLOATING BOX
BBC MUSIC CENTRE

FOREIGN OFFICE ARCHITECTS
LONDON, 2003 (competition)
CLIENT: BBC

FOA's visualisation of the building

In order to secure the competition we realised that we had to prove it would be very easy to build this scheme, to show how simple the engineering was. The focus of our structural contribution was the Studio building, which was an exercise in making a very compact space out of the two large performing studios.

We got involved with FOA right from the beginning of the concept. Because the driving forces are mostly to do with acoustics, the first question was to determine how to float an acoustic box inside the space. As they started to work from the inside out and putting it on the site, they presented the idea of the rolling films as a metaphor that would relate to the history of the BBC, giving some direction in terms of the image of the project. The way we had to constrain it — because both cost and site were significantly constrained — was by making most of the transverse direction of the building very opaque for acoustic reasons, as well as for visual considerations, and creating a very large window at the front of the building. FOA's "ideal" position, which wasn't so common in those days, was to use transparent glass that would also have the high-quality acoustic performance demanded by the studios. But they also wanted to create transparency as a theme for an institution that is predominantly opaque and that is only partly exposed through media. They wanted to enact that transparency on the site, to enhance the experience of people being able to connect from the outside to what was going on inside (the orchestra here is one of the leading ones in the UK) but also for the people inside to be able to see outwards. It all became a game between two kinds of extremes in terms of totally opaque in one direction and totally transparent in the other – opaque in acoustic terms and transparent in visual terms.

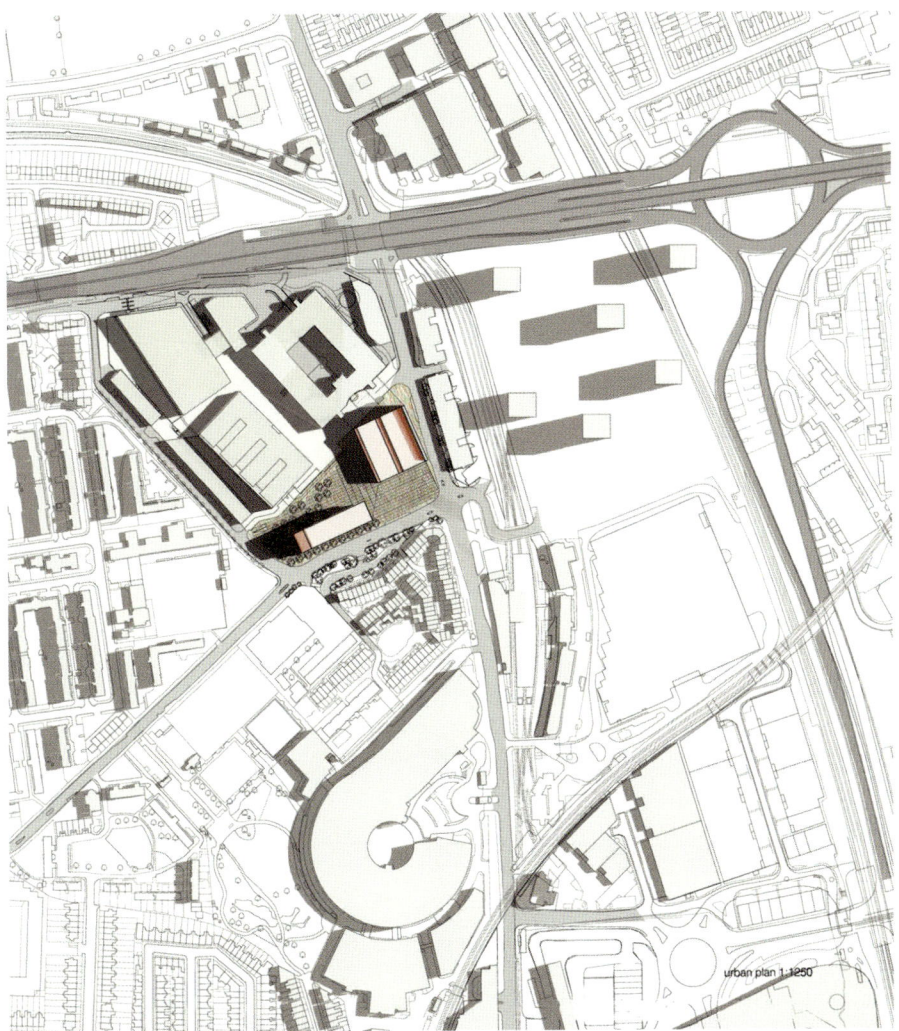

urban plan 1:1250

The shape of the building is defined by a wrapping surface under and over the studio and which contains public space and a cafeteria, but this surface also grows above the building in the form of a big projection screen. Therefore the surface also became an event space projecting what was going on inside to the outside. In this sense, the rolling film metaphor was also conditioned by the fact that the site is scarred on one side by a very big elevated motorway from which these projections would be perceived.

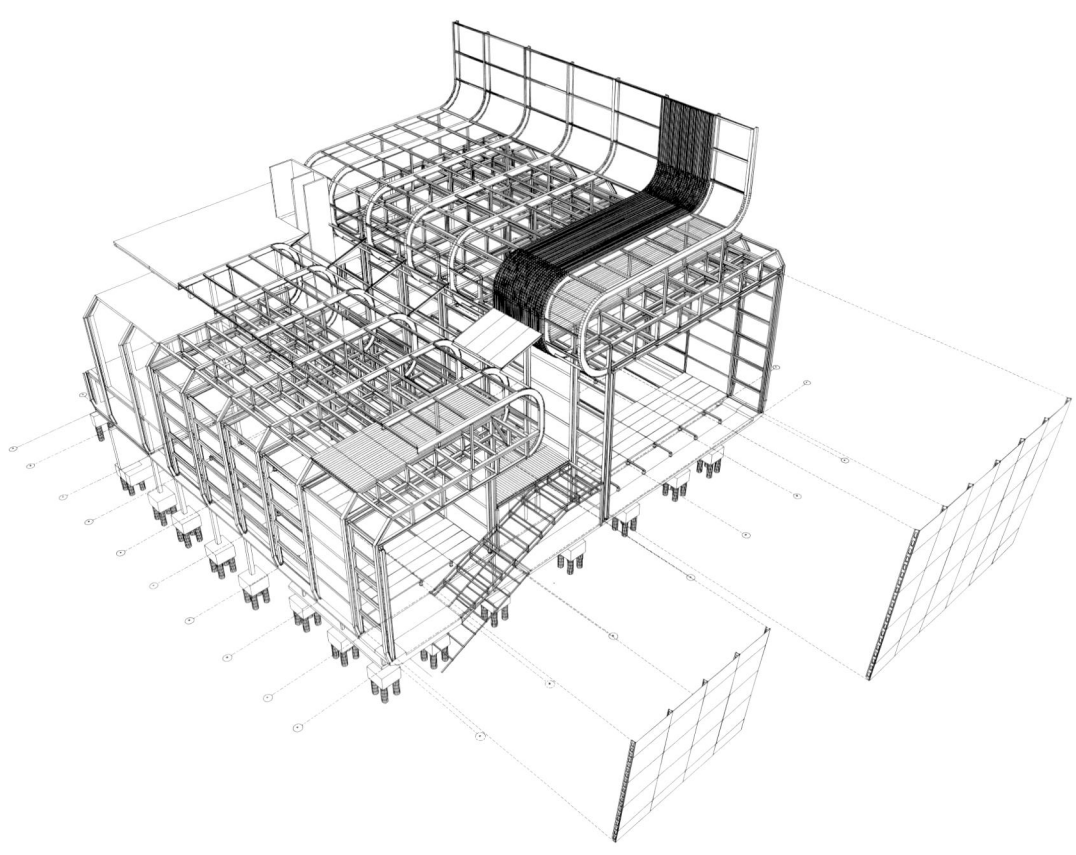

The structural simplicity

To make all of those features possible within the budget limitations, the structure had to be very simple. What you see in our structural diagrams is really about convincing the client that this is the most compact way in which to wrap a building. And the fact is that most of the function of the structure is to create large volumes with clear, free space. Although the initial images suggested the wrapping surface would be made out of concrete. We suggested that the best way to construct a floating box inside a big envelope was to create it in lightweight steelwork, which would also allow it to be assembled with any materials that would support different skin finishes or applications. So we kept the structure as simple as possible in order to get the other parts to be affordable.

Then we extruded the scheme and planned a very simple construction around that. At the back of the building, which houses the actual work spaces, you will see from the renderings that there is a concrete block coming up at the back that gives stability to the building, so that the main floating volumes are basically propped against the rear of the building. You get a ground plane which is concrete at the entrance foyer, and then that concrete wraps up at the back to act as a stable cellular space. Once you create that structure as a sort of 'drawer', the whole frame sits as the 'contents' of the drawer in a very simple way.

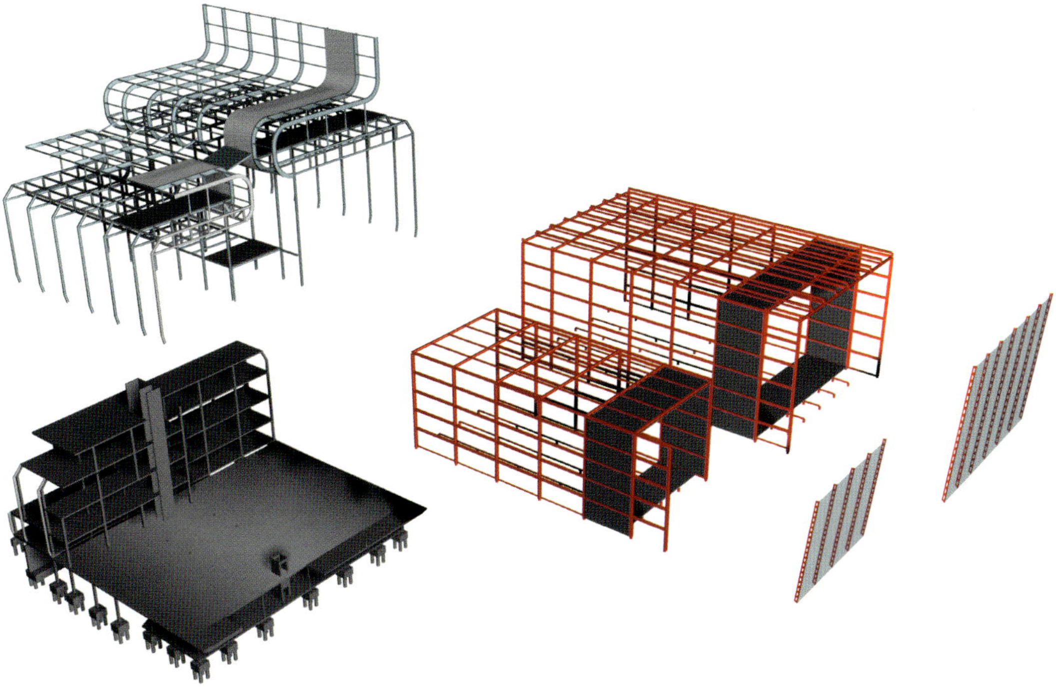

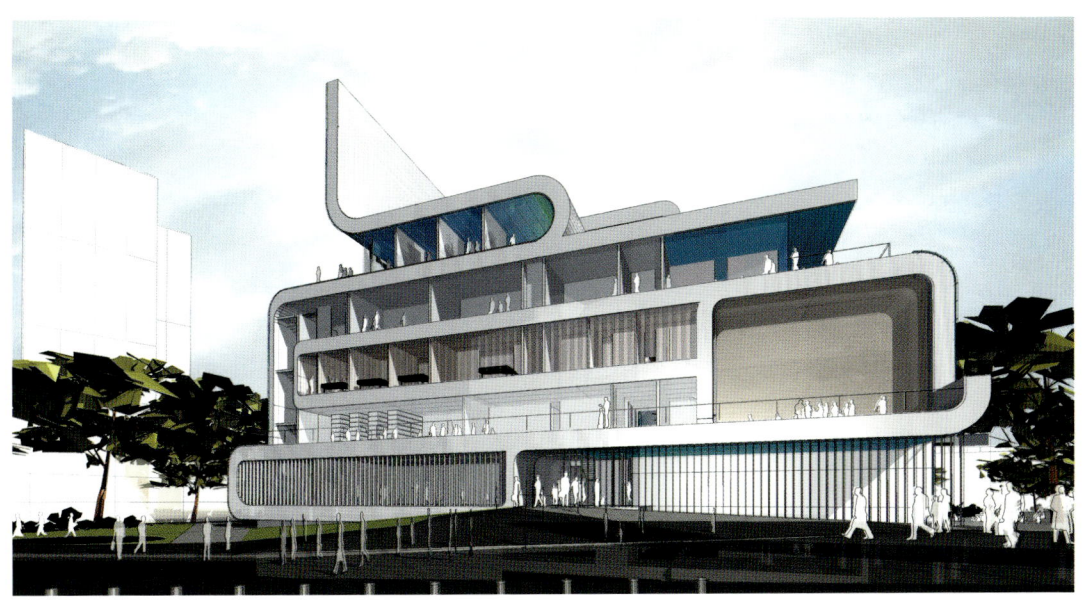

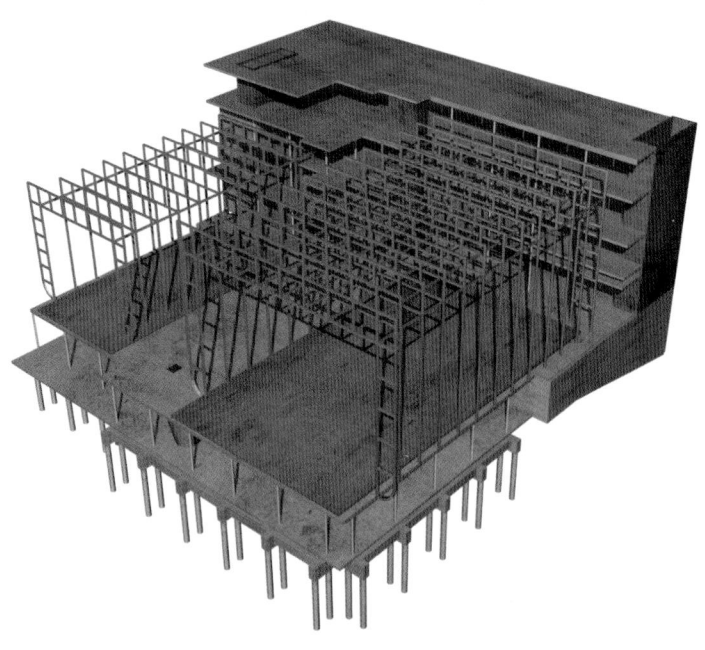

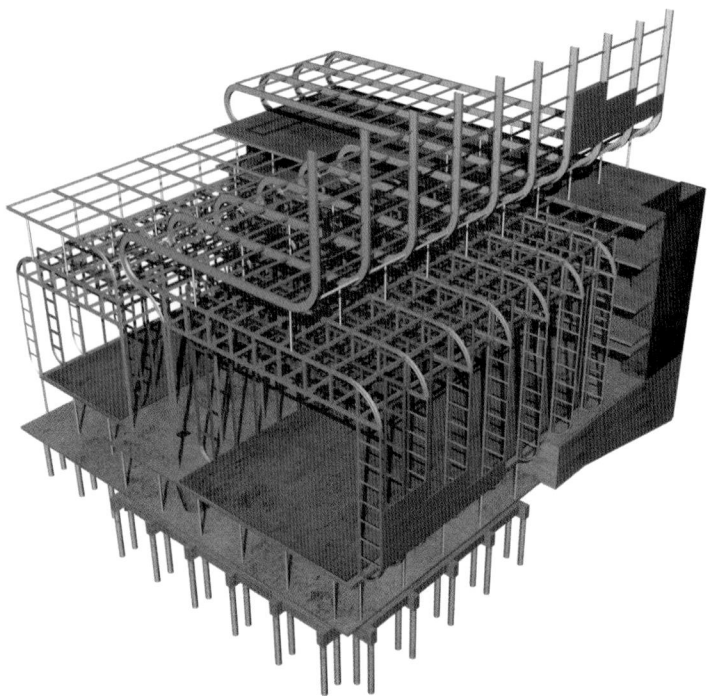

Structural Isonometrics

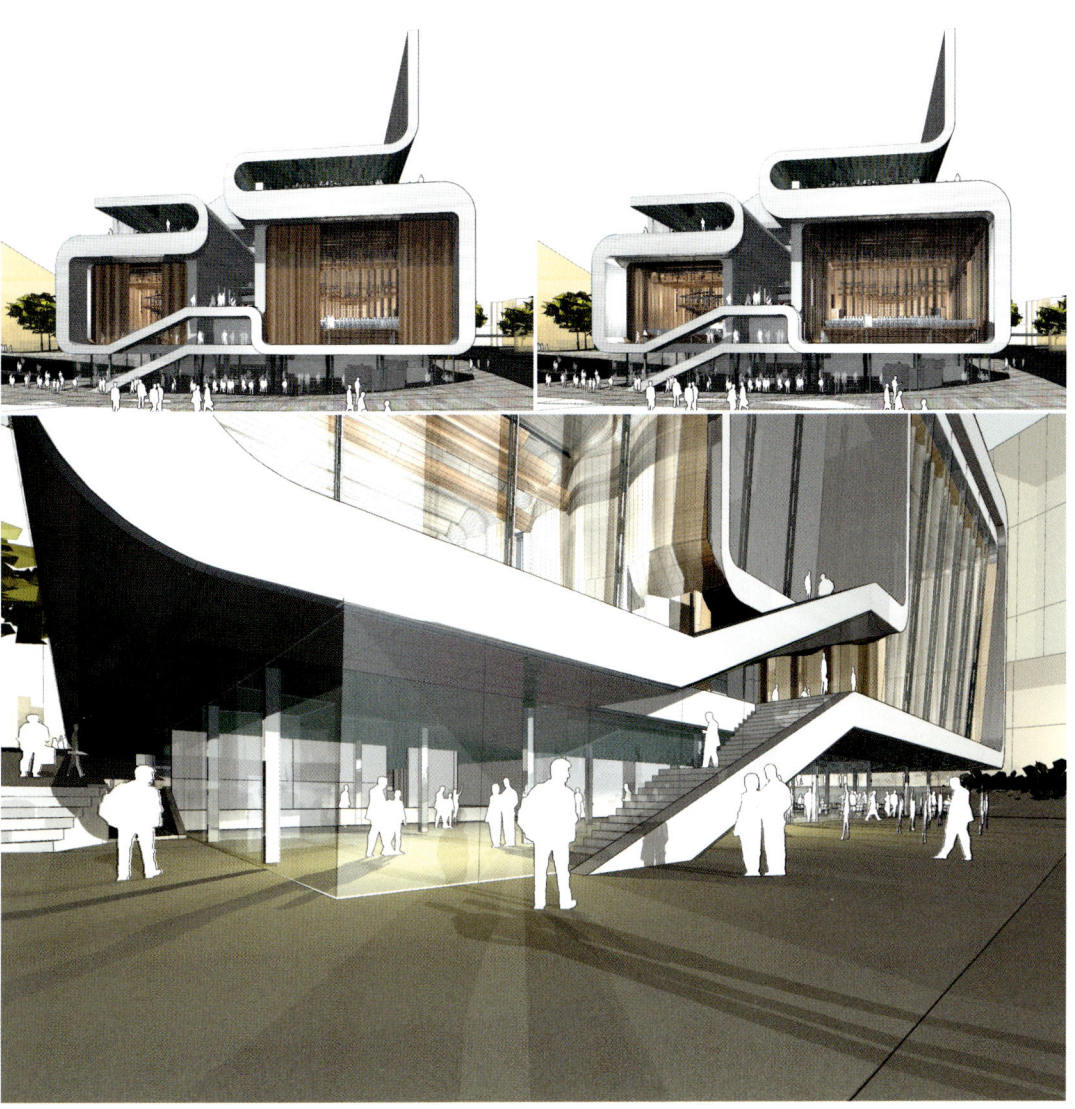

We had to develop this simple organisation because we knew there would be a lot of questions about the other technologies that were being introduced, such as the two layers of large sheets of glass that form the transparent façade, and their acoustic and structural behaviour under different sound frequencies. So in order to have them buy most of that extra technology, the internal structural system became very simple. Also, FOA's decision to place the two studios side by side instead of stacking them is something no other entry did. This adjacency defined an intermediate buffer zone which was used as an entry and relational space, contributing to the compacity of the whole. And we won the competition and proved it was all possible, but it is still on hold.

FOA were invited to take part in an exhibition in Malmö for which they had to design a 'house' that could be transported by rocket and deployed in Mars.

The design developed by FOA relies on splitting a cylinder once it has been transported to site. The transformation of such a geometry results in two interlocking cylinders in the unfolded form, where the surface area increases by 100%.

The structure is based on very simple circular ribs that can be rotated, while the covering required is a 'fabric' that can expand and contract to this extent. Additionally this surface would need to deal with the harsh temperature changes one would encounter in this environment. The project remains unrealised.

FABRIC FRAME FUTURE HOMES– MARS HOUSE

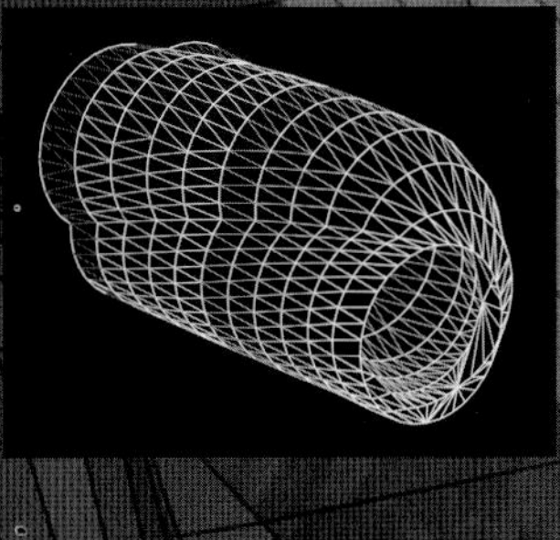

FOREIGN OFFICE ARCHITECTS
TRAVELING EXHIBITION, 2000
CLIENT: CONFIDENTIAL

	REGION 1	REGION 2	REGION 3	REGION 4	REGION 5
PANEL 1					
PANEL 2					
PANEL 3					

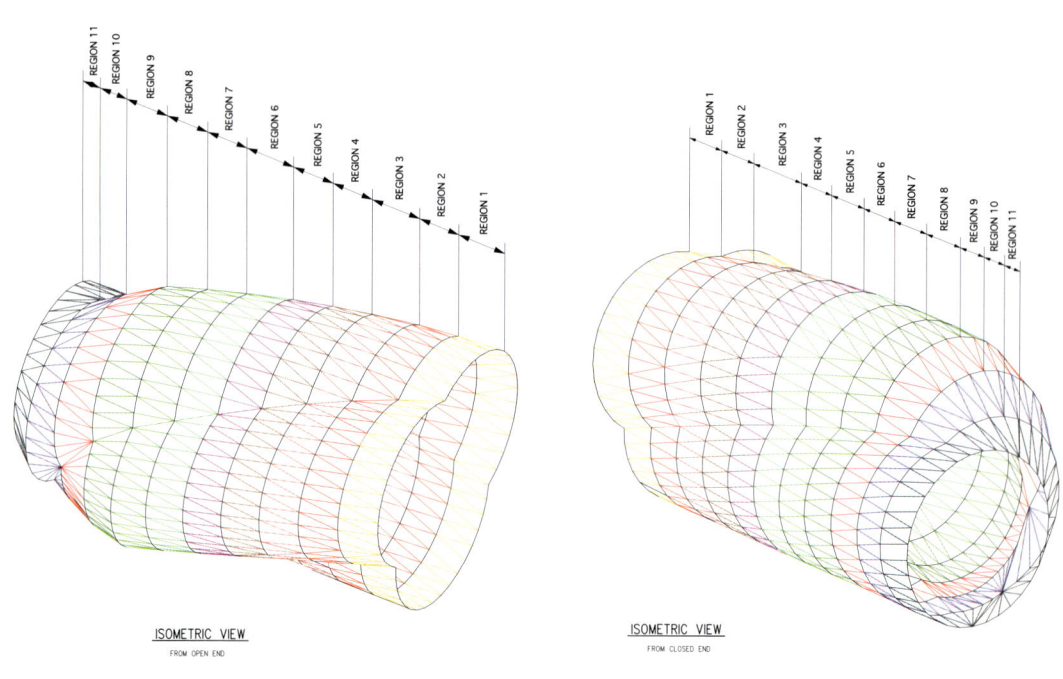

ISOMETRIC VIEW
FROM OPEN END

ISOMETRIC VIEW
FROM CLOSED END

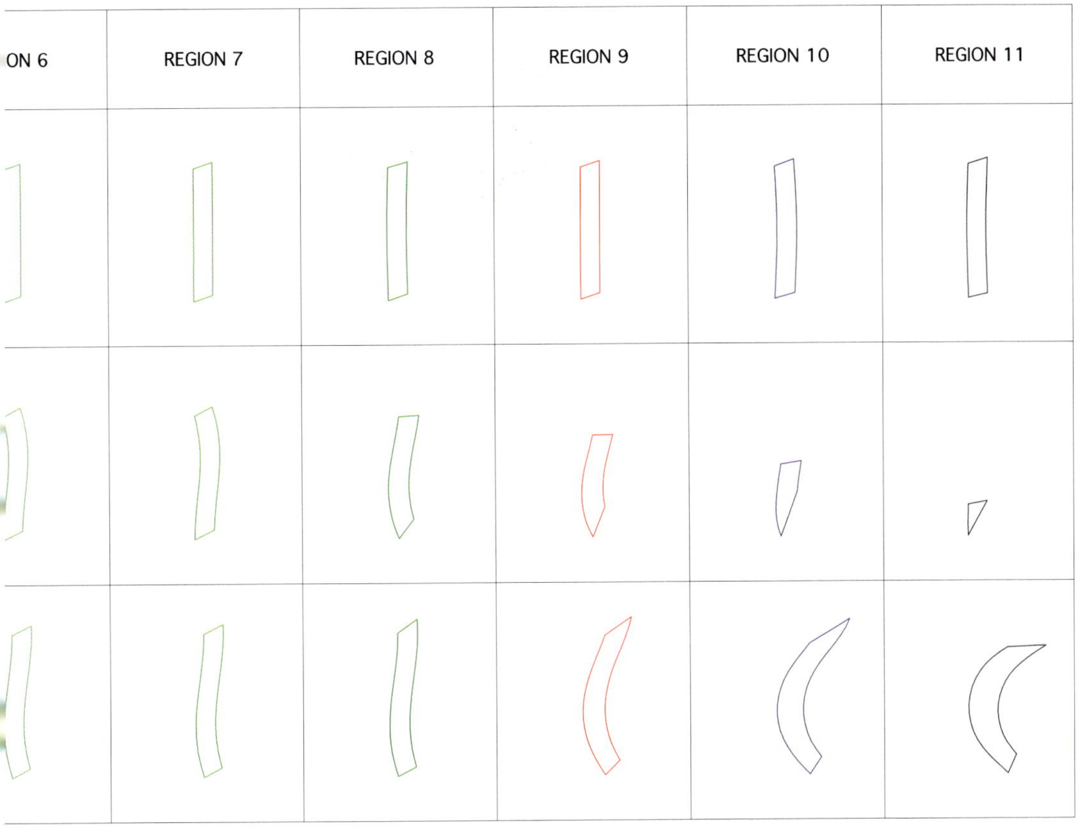

When we did the Mars House – the cylinder that opens – we created a drawing (in collaboration with Tensys) of how to un-tile something when it expands by 100 percent. (To produce it, we actually un-tiled the fabric into certain panels.) In terms of representation, it illustrates how engineers communicate certain ideas and strategies. Some projects run the risk of being dismissed on the basis that they are impossible to build, so during the presentation stage, we wanted to show how simple the engineering was and the only way to communicate that was to do a self-explanatory 3-D construction drawing. Representation is extremely important at certain stages.

The Mars house project was a challenge because we never found a fabric that could expand by so much and still retain its integrity. A bungee rope is the only thing that can go through 100% expansion and recover, but it could not withstand the 1000°C temperatures on Mars. There were thoughts that you could invent the material to do this in this future.

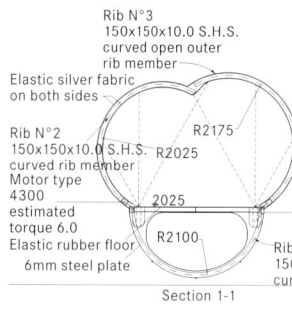

Rib N°3
150x150x10.0 S.H.S.
curved open outer
rib member

Elastic silver fabric
on both sides

Rib N°2
150x150x10.0 S.H.S.
curved rib member
Motor type
4300
estimated
torque 6.0
Elastic rubber floor
6mm steel plate

R2175

R2025

2025

R2100

Rib N°1
150x150x10.0 S.H.S.
curved rib member

Section 1-1

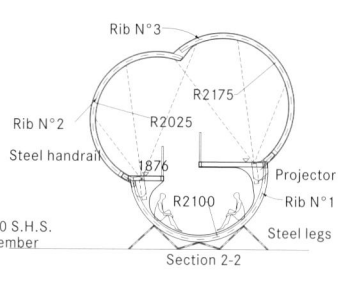

Rib N°3

Rib N°2

Steel handrail

R2175

R2025

1876

R2100

Projector

Rib N°1

Steel legs

Section 2-2

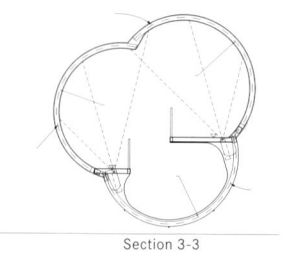

Section 3-3

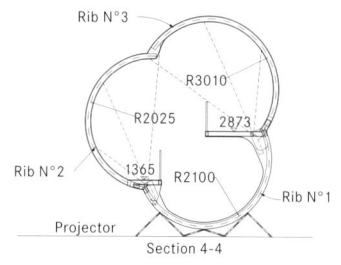

Rib N°3

R3010

R2025

2873

Rib N°2

1365

R2100

Rib N°1

Projector

Section 4-4

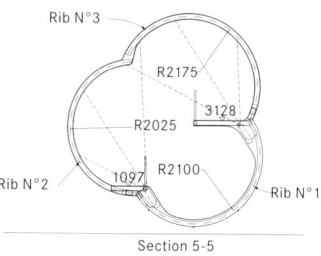

Rib N°3

R2175

R2025

3128

R2100

Rib N°2

1097

Rib N°1

Section 5-5

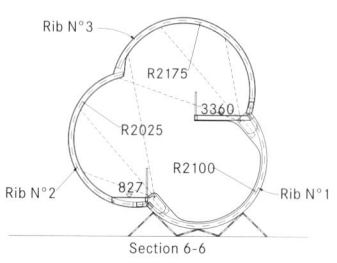

Rib N°3

R2175

3360

R2025

R2100

Rib N°2

827

Rib N°1

Section 6-6

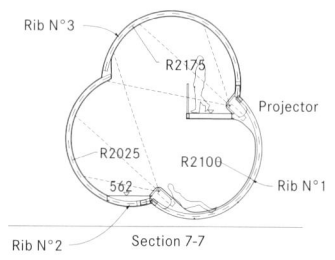

Rib N°3

R2175

R2025

R2100

562

Rib N°1

Rib N°2

Section 7-7

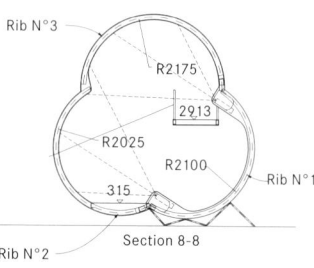

Rib N°3

R2175

2913

R2025

R2100

Rib N°1

315

Rib N°2

Section 8-8

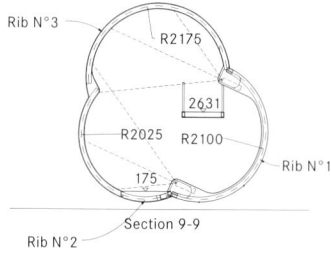

Rib N°3

R2175

2631

R2025

R2100

Rib N°1

175

Rib N°2

Section 9-9

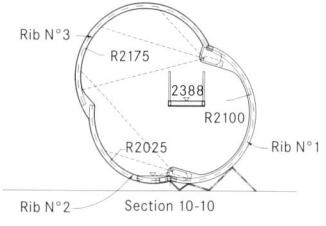

Rib N°3

R2175

2388

R2100

R2025

Rib N°1

Rib N°2

Section 10-10

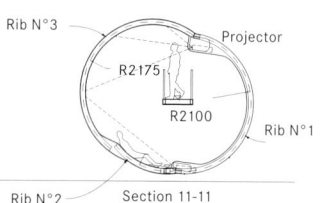

Rib N°3

Projector

R2175

R2100

Rib N°1

Rib N°2

Section 11-11

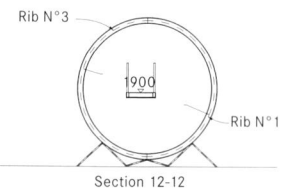

Rib N°3

1900

Rib N°1

Section 12-12

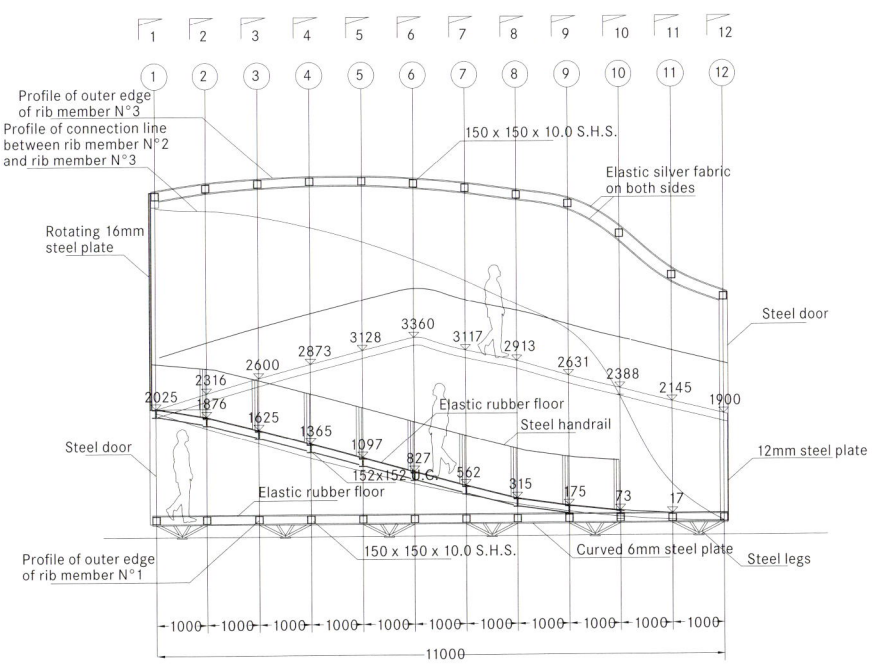

FOA's visualisation of interiors

Profile of outer edge
of rib member N°3
Profile of connection line
between rib member N°2
and rib member N°3

150 x 150 x 10.0 S.H.S.

Elastic silver fabric
on both sides

Rotating 16mm
steel plate

Steel door

3360
3128   3117 2913
2873   2631
2600   2388
2316   2145
2025   1900
1876
1625
1365
1097
827
152x152 U.C.
562
315
175
73
17

Elastic rubber floor

Steel handrail

12mm steel plate

Steel door

Elastic rubber floor

Profile of outer edge
of rib member N°1

150 x 150 x 10.0 S.H.S.

Curved 6mm steel plate

Steel legs

1000 1000 1000 1000 1000 1000 1000 1000 1000 1000 1000

11000

AKT collaborated with Zaha Hadid Architects on a 200 metre-high commercial tower in Milan's Fiera district. Flanked by neighbouring designs by Arata Isozaki and Daniel Libeskind, the skyscraper uses a tangential twist to generate curves with straight lines. Geometrically, each floor shifts slightly in relation to the ones below and above it. Vertical columns, per conventional structural thinking, are inefficient because they are in a different position on each floor relative to the core and perimeter, and require a transfer beam. AKT's structural solution proposes two pairs of subtly curving columns, which read as an integral part of the overall form, and also help to establish a rhomboidal pattern for the glazing on the façade.

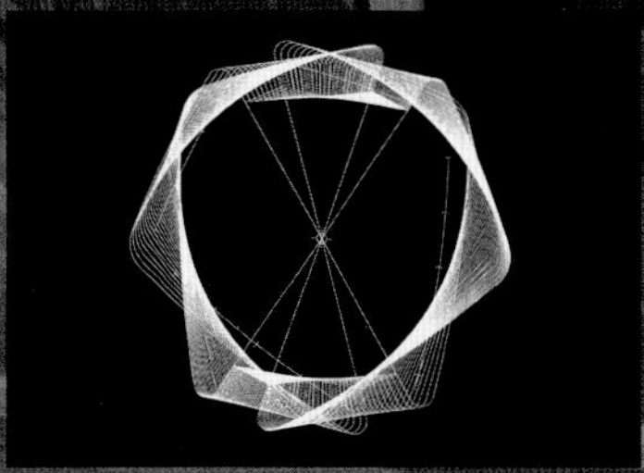

TWISTING SKIN, REGULAR STRUCTURE
MILAN FAIR TOWER

ZAHA HADID ARCHITECTS
MILAN, ITALY, 2012
CLIENT: CITYLIFE

Visualisation courtesy of client and Zaha Hadid Architects

This tower, when read with the two adjacent towers, demonstrates the potential of extreme types in what is the structural engineers classic prototype – the tall building.

At first sight, the geometry of the tangential twist on plan suggests that the whole structural system needs to be warped. The twist reduces as one ascends the tower, the formal logic of this being that the views of the City of Milan change with height. The 'slots' on the floor plan allow this rotating experience but, in addition, we were able to take advantage of this slot by inserting sheer walls on each site.

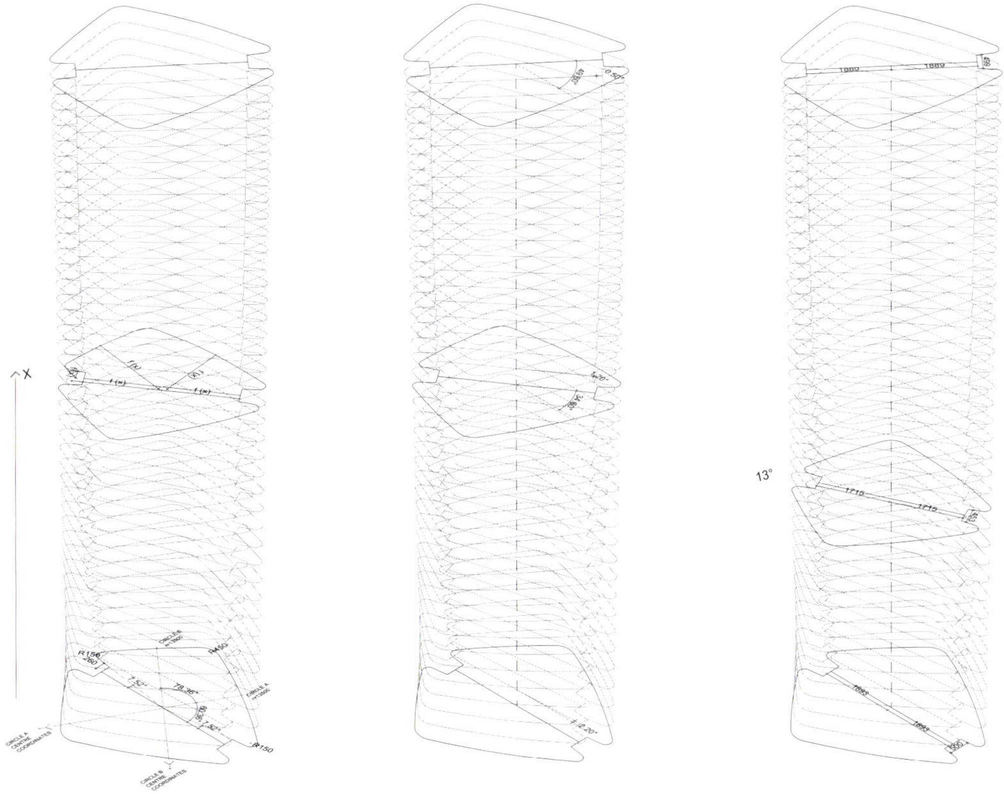

As the floor plates dimension varies from up to 14m, there was a temptation to insert columns between the core and external columns to free the skin. Instead, after closer examination of the geometry, we chose to push all the columns to the perimeter. This shift would inevitably mean that the size and geometry (avoid twisting) of the columns would need careful assessment. This arrangement of tilted columns then includes a built-in twist in the core structure that had to be resisted.

To coordinate the façade with the column arrangement a number of patterning studies were necessary, the pre-requisite being that a standard, repetitive flat panel of glass was to be used. The properties of the glass panel and its limit in terms of "bending flat-panels" had to be established.

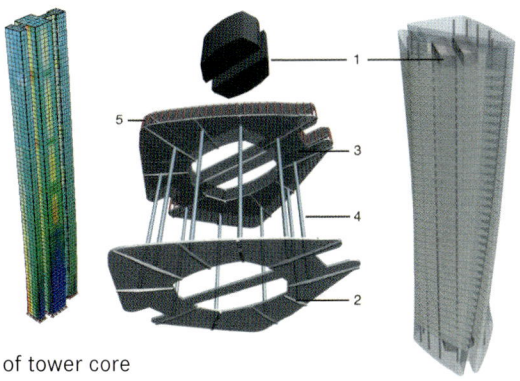

Stress distribution of tower core

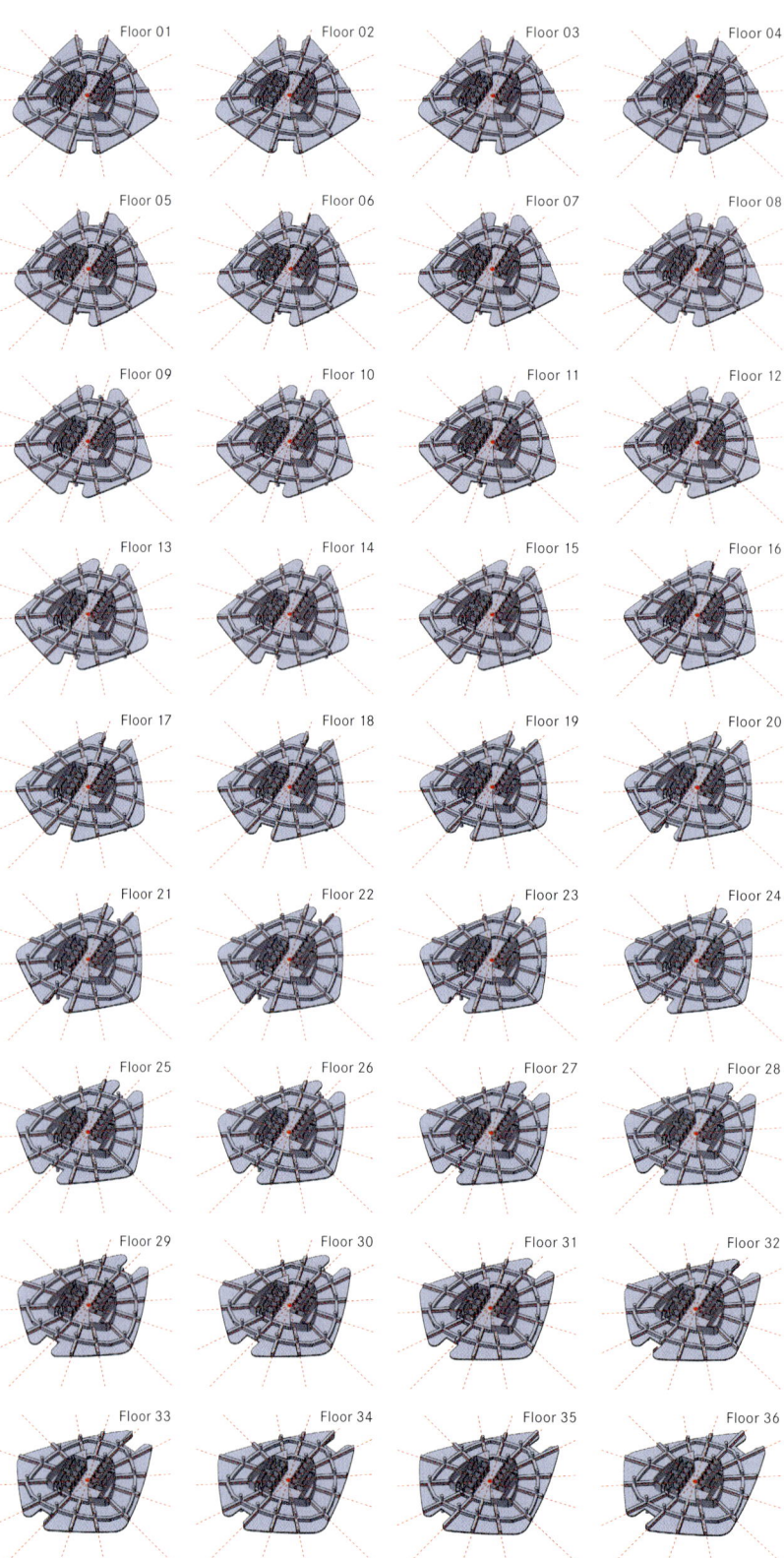

Floor 01 Floor 02 Floor 03 Floor 04
Floor 05 Floor 06 Floor 07 Floor 08
Floor 09 Floor 10 Floor 11 Floor 12
Floor 13 Floor 14 Floor 15 Floor 16
Floor 17 Floor 18 Floor 19 Floor 20
Floor 21 Floor 22 Floor 23 Floor 24
Floor 25 Floor 26 Floor 27 Floor 28
Floor 29 Floor 30 Floor 31 Floor 32
Floor 33 Floor 34 Floor 35 Floor 36

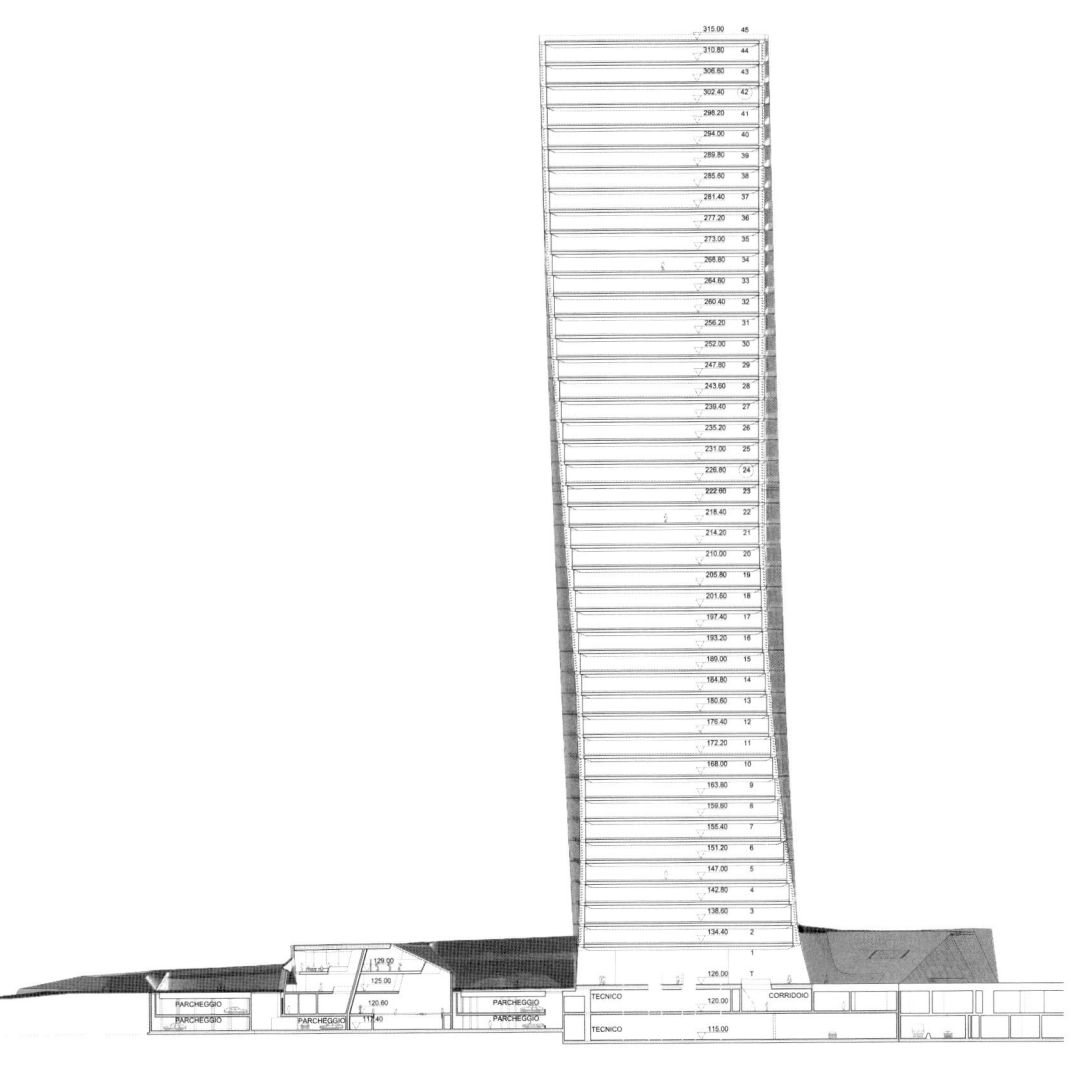

We refrained, in this case, from introducing any structural feats and worked with a conventional system where the core provides the lateral stability and resists secondary forces that are generated from the 'twist' in the columns. To achieve this the floors, which support vertical loads, have an additional role of transferring torsional forces generated by the columns.

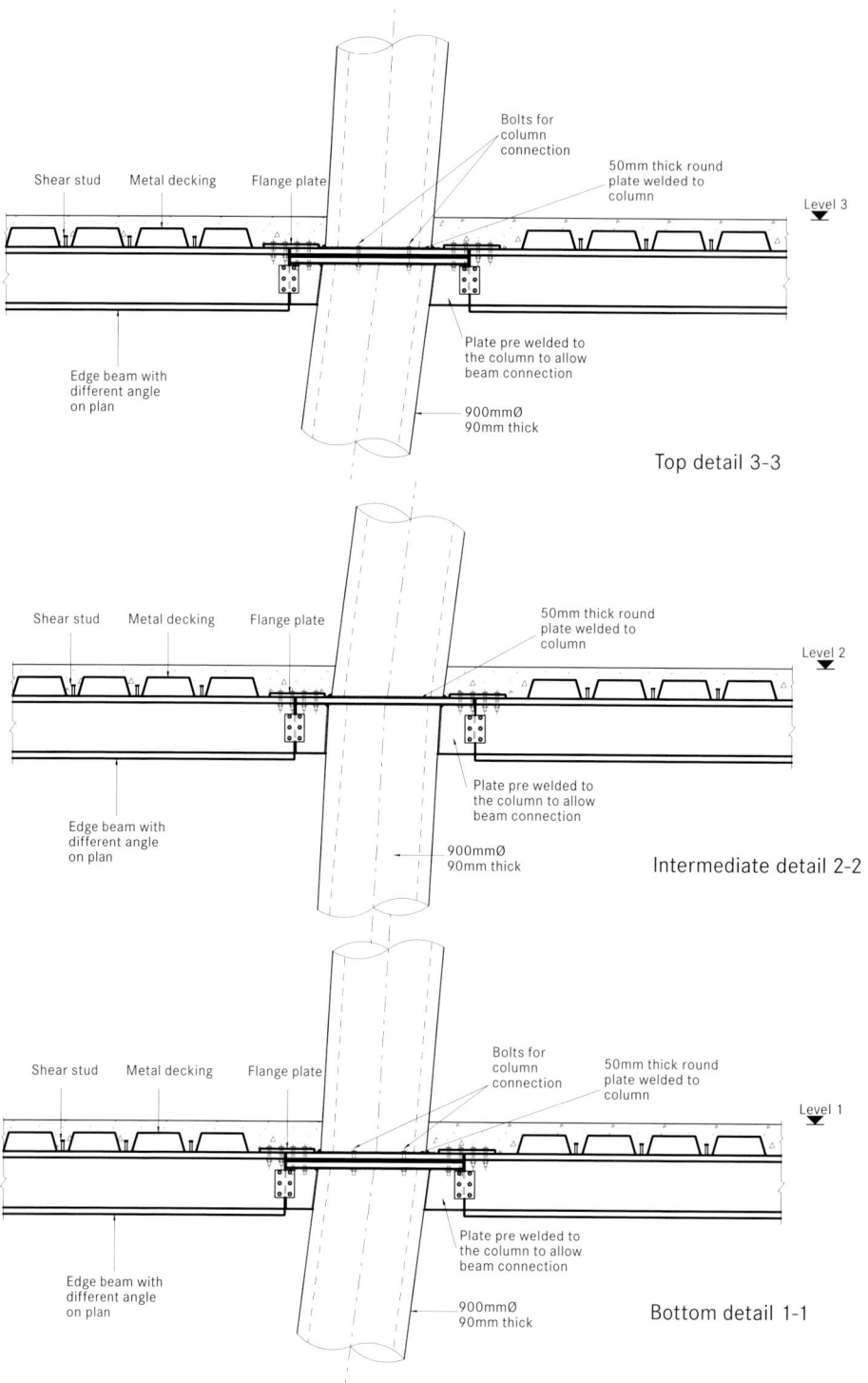

Shear stud   Metal decking   Flange plate

Bolts for
column
connection

50mm thick round
plate welded to
column

Level 3

Plate pre welded to
the column to allow
beam connection

Edge beam with
different angle
on plan

900mmØ
90mm thick

Top detail 3-3

Shear stud   Metal decking   Flange plate

50mm thick round
plate welded to
column

Level 2

Plate pre welded to
the column to allow
beam connection

Edge beam with
different angle
on plan

900mmØ
90mm thick

Intermediate detail 2-2

Shear stud   Metal decking   Flange plate

Bolts for
column
connection

50mm thick round
plate welded to
column

Level 1

Plate pre welded to
the column to allow
beam connection

Edge beam with
different angle
on plan

900mmØ
90mm thick

Bottom detail 1-1

2 storey high - steel circular column (the H-section with plate is the same)

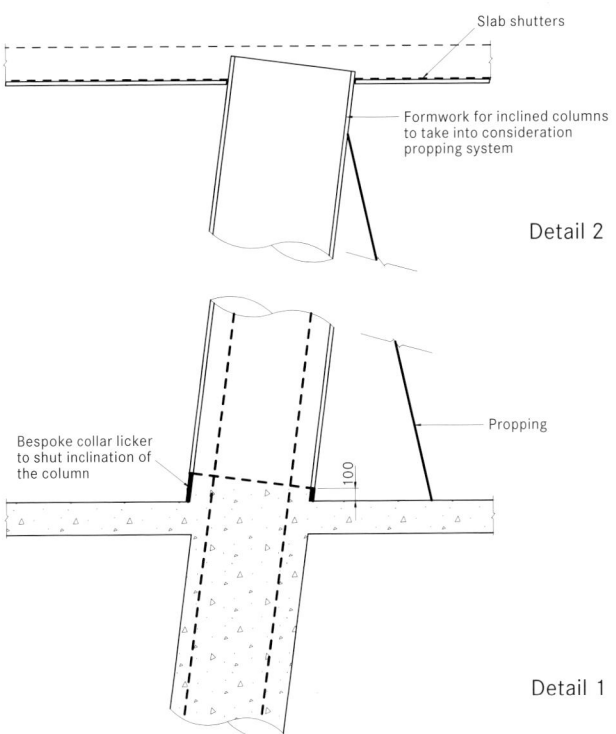

Slab shutters

Formwork for inclined columns
to take into consideration
propping system

Detail 2

Propping

Bespoke collar licker
to shut inclination of
the column

100

Detail 1

Construction details for RC concrete

In order to integrate the columns with the façade we have been looking at both reinforced concrete and steel – and in fact, the project has allowed us to consider the latest development in both steel and concrete as primary materials for tall buildings. The columns need to reduce in size as we get higher and this change of dimension affects the geometry and the aesthetics. They are likely to use the highest strength of concrete (200-250n/m^2 similar in strength to low grade steel). The tilted in-situ construction is developed by only varying the kicker so that the formwork can be reused, as a way to overcome the fact that every angle is different and we didn't want to use new shutters each time. In addition, at the other extreme, we have been exploring steel columns with super high strength steel of 500n/m^2.

A competition for what would have been, at 575m, among the tallest buildings in the world gave AKT and Foreign Office Architects an opportunity to explore how the different logics of structure, use and value could interact to create a new architectural paradigm for ultra tall towers. Basically the design consists of three rising segments which transform as they rise into different generic floorplates, and the overall structure follows that change of shape.

At low level the shape is triangular, which establishes a structurally stable perimeter. It also gives deep plan space, suitable for the retail and commercial uses which are best located lower in a tall building. As the building rises the perimeter steps in, giving more external wall-to-floor area which better suits speciality offices and residential uses. Plant areas are combined with the transfer structures at the levels where the shape changes, minimising the loss of lettable or saleable space.

Its structure is a hybrid of very high strength concrete core with steel columns and outriggers on the perimeter, while overall shape also helps to offset the effect of both earthquakes and wind load. Overall, the aim is to merge the concept of structural efficiency with architecture in the rarified context of an ultra tall building. It shows how the most stable structural form can also give the best commercial outcome.

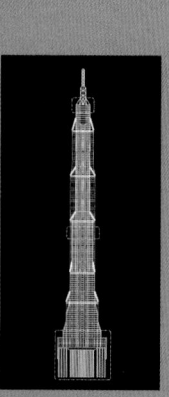

SUPER
HIGH-RISE
BUSAN
TOWER

FOREIGN OFFICE ARCHITECTS
BUSAN, SOUTH KOREA, 2006
CLIENT: SOLOMON GROUP

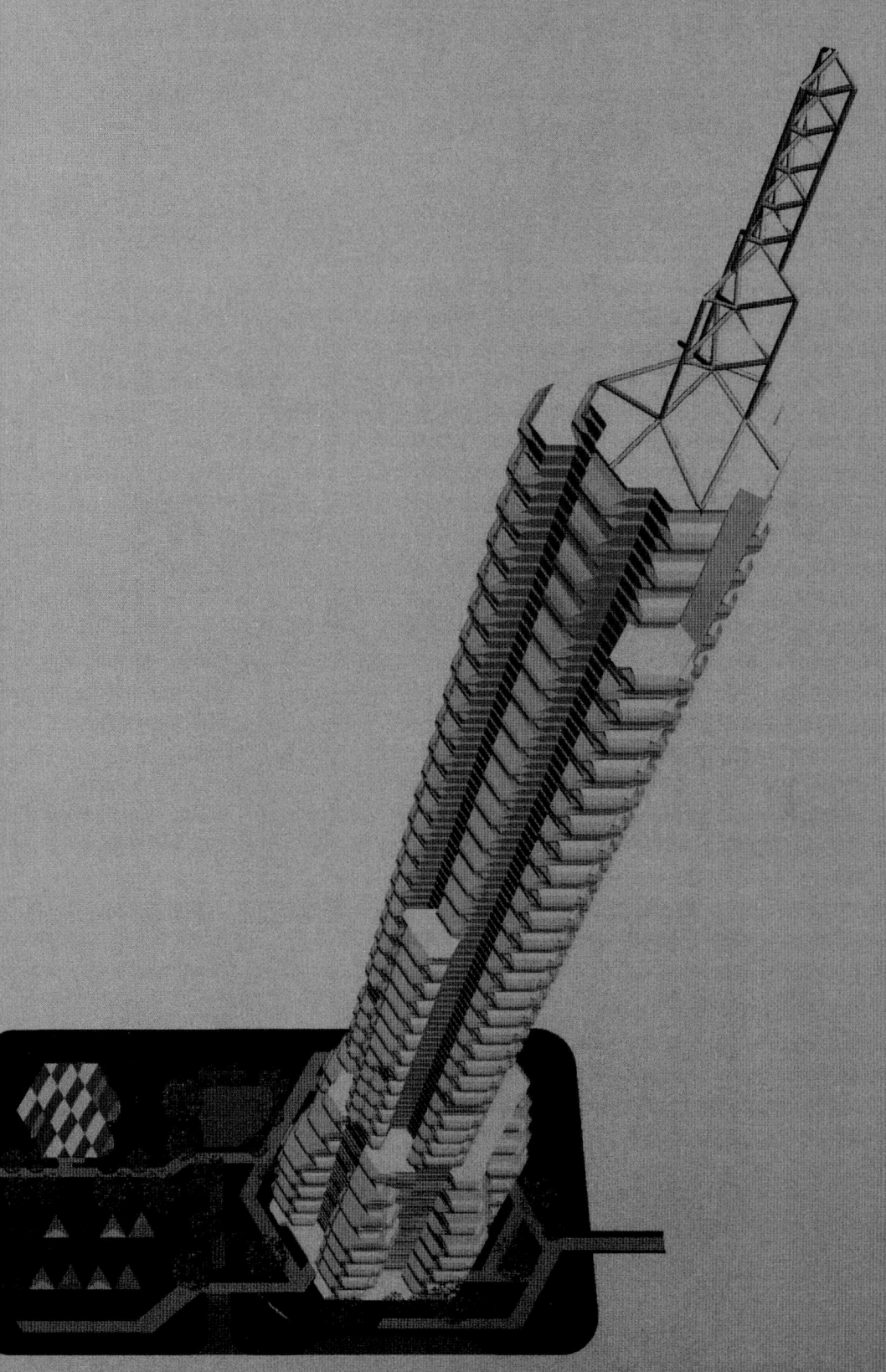

Would you say that the ultimate test for engineers is still the high-rise?

If one is considering the singular role of structural engineers, then there is no question that the programmes that are perceived as posing the most extreme challenges are the high-rise that cantilevers from the ground and the bridges as it spans a distance. The tower in particular has re-entered the imagination of engineers recently, perhaps even more so than during the time when Fazlur Khan and his colleagues invented great models.

What's peculiar about the Busan project is the geography (the cultural context of technology and construction in South Korea) as well as the bold approach in the competition to invite contemporary architects to re-think the high-rise. How do you fit into this?

Well, FOA invited us to their team in an equally 'bold' move, as the normal process would be to approach engineers who had already built towers. Our first response was to look at the panorama of world towers and zoom into Korea in an attempt to establish what and how high-rises have already been constructed there. Standing at a height of 550m, this would be one of the tallest. The technology is mostly available, and the real opportunities to develop projects in this way lie in fast-growing countries such as Korea.

How did you approach the structure?

FOA had already used the metaphor of 'snowflake like' plans that optimise the relationship between façade and floor area and assist structural strategies. Taking a lead from these plans we made simple intuitive analysis of structural forms on plan and elevation – for instance that a square plan (four corners) departs four primary loads compared to a triangular plan (3 corners), or that elevations that become lighter with height are more efficient as the base carries most load. These are simple intuitions about extremes that we looked into. We then understood that a plan with 'eight corners' allows many more structural possibilities and, more importantly, improved floor-to-perimeter ratios that calibrate a higher order of efficiency.

World towers

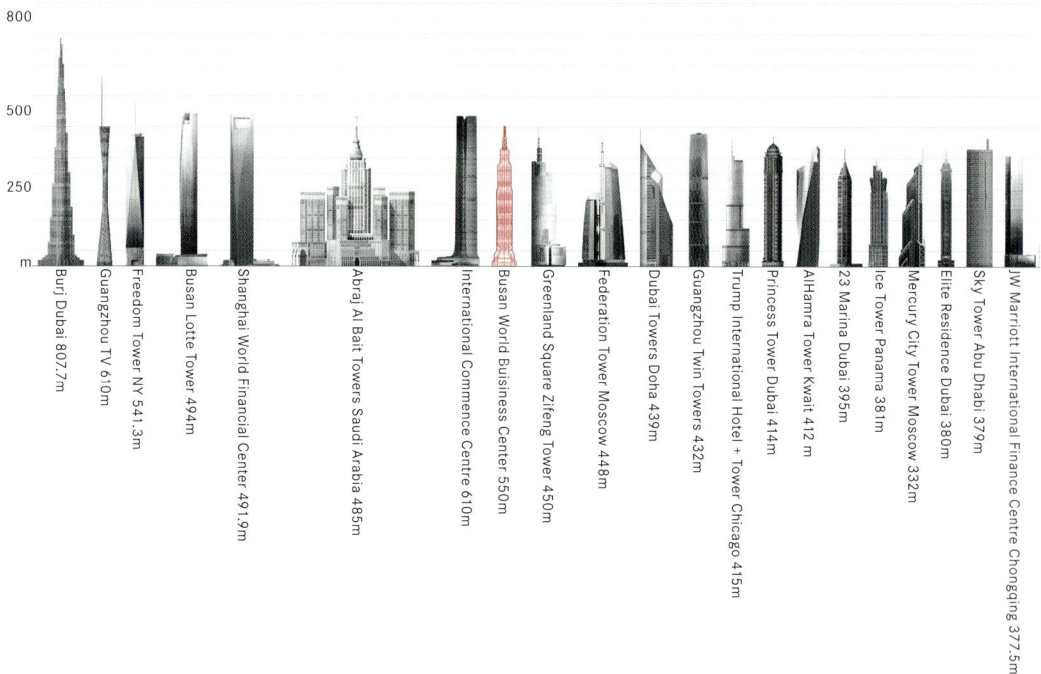

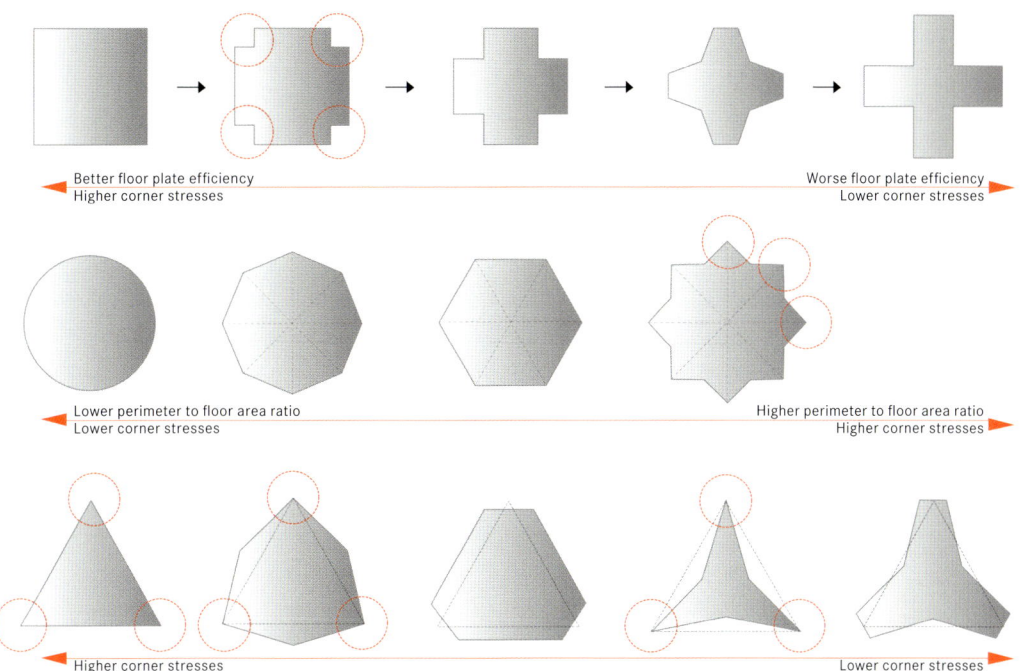

Better floor plate efficiency
Higher corner stresses

Worse floor plate efficiency
Lower corner stresses

Lower perimeter to floor area ratio
Lower corner stresses

Higher perimeter to floor area ratio
Higher corner stresses

Higher corner stresses

Lower corner stresses

Plan analysis

Elevation analysis

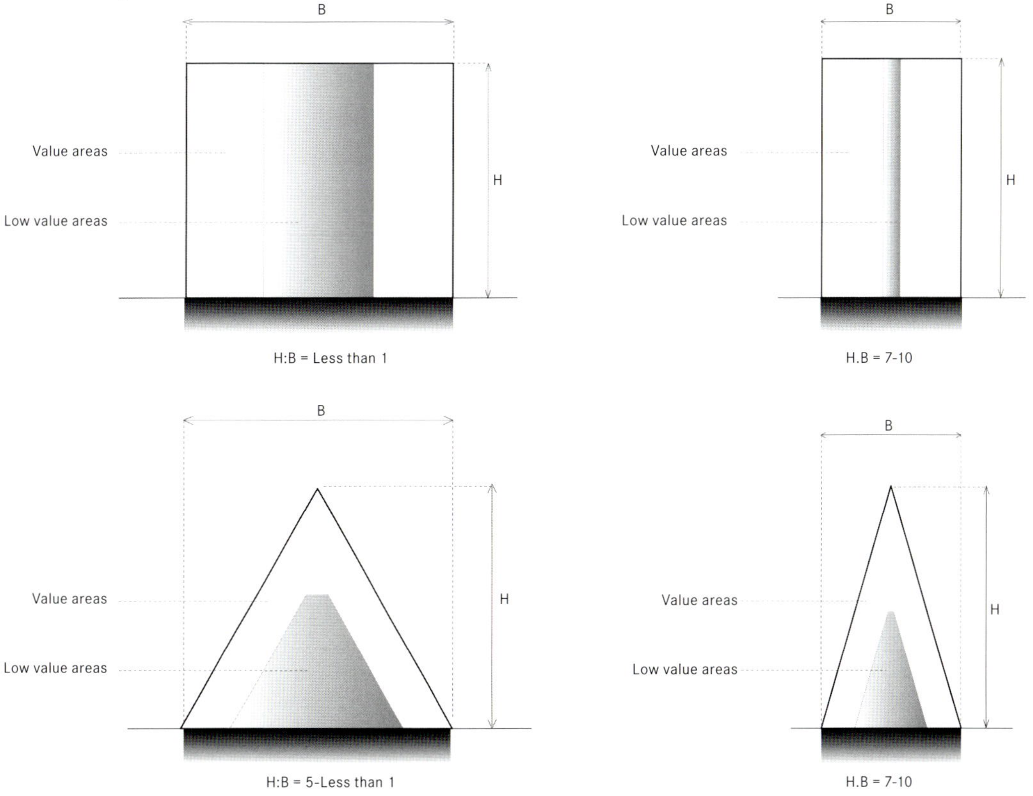

B

B

Value areas

Low value areas

H

Value areas

Low value areas

H

H:B = Less than 1

H.B = 7-10

B

B

Value areas

Low value areas

H

Value areas

Low value areas

H

H:B = 5-Less than 1

H.B = 7-10

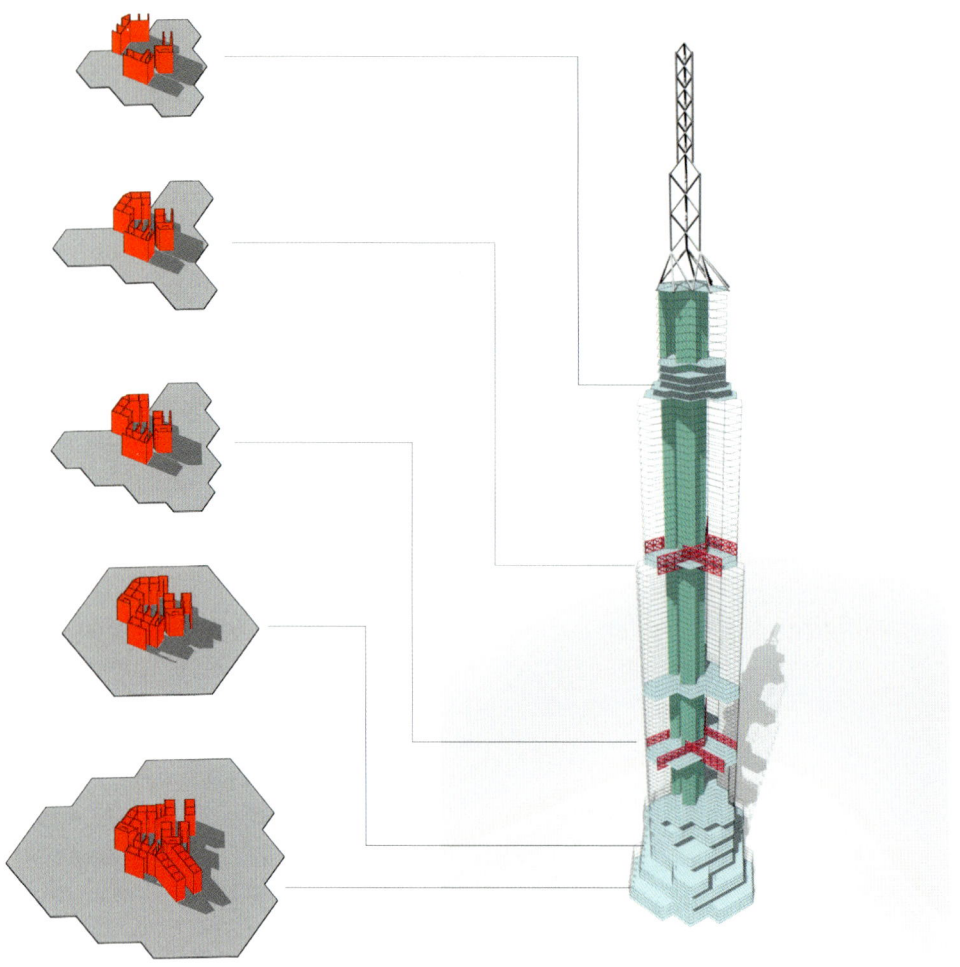

There are clear limits at such extremes that are already tried and tested, so the room to re-invent is limited. Most innovations therefore come from reshaping known forms and ensuring that they will work. FOA had, in their previous work, conducted a lot of research in arriving at a hexagonal plan for the core. In this project, on the other hand, the typhoon wind and seismic conditions make the lateral stability considerations even more important. So by shaping the core and combining it with the external façades through floor outriggers, stability can be achieved for this form. This hybrid of 'soft tube' has the benefit that the façade becomes simpler, compared to the external 'tubes' which are structurally more efficient.

How to cite this was the input of the client at this stage, and did this have an affect on the process at all?

As we discussed earlier, in these parts of the world speed is essential. The client was also very clear that the use of reinforced concrete must be maximised. We looked into the latest developments in concrete and it has to be said that the Far East, particularly Japan, have led the world in the use of high-strength concretes. (Normal concrete strength is $40n/m^2$, whereas very high-strength concrete is $250n/m^2$). So, particularly for the cores at the base that support the highest stress, $100n/m^2$ strength concrete was proposed. The columns had to be as small as possible so we looked at a variety of steel columns.

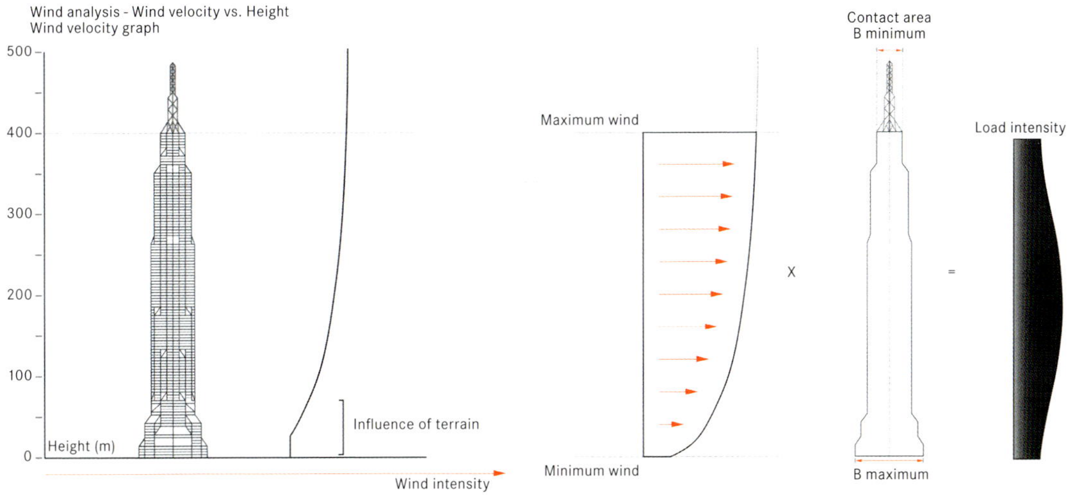

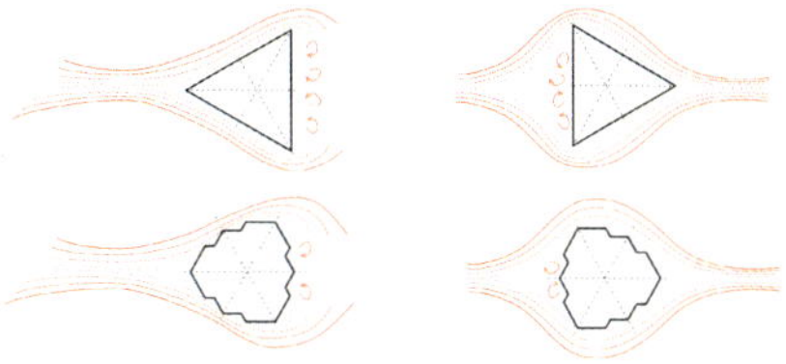

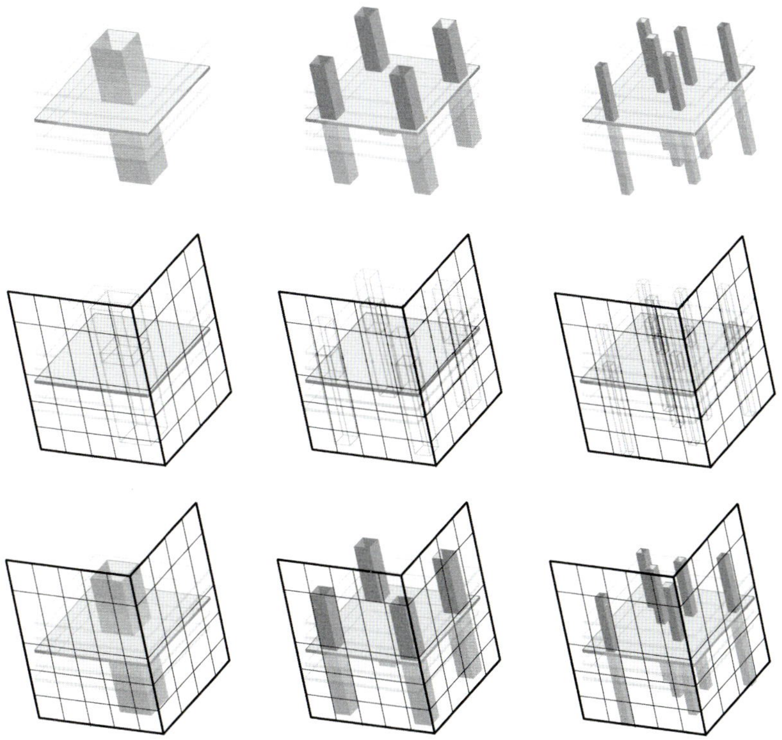

Wind analysis - Wind velocity vs. Height
Wind velocity graph

500 —

400 —

300 —

200 —

100 —

0 —

Height (m)

Influence of terrain

Wind intensity

Contact area
B minimum

Maximum wind

X

=

Load intensity

Minimum wind

B maximum

Stability derived from
circulated cores

Stability derived from
external 'tube' system

Hybrid stability system

Central core

Perimeter core

Multiple cores

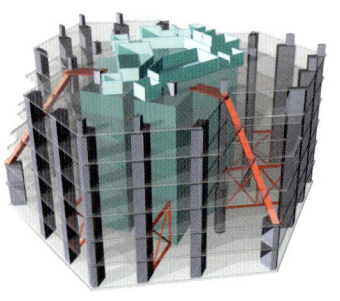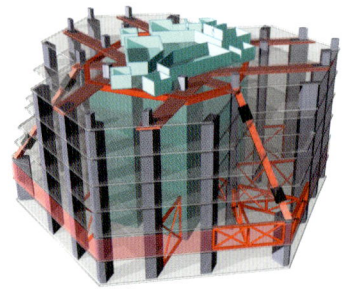

Would you say FOA's explanation of the tower would be the same as yours?

That is the point; their approach and starting position means that they initiate and understand these issues and then 'play' with our responses. We would thus exploit a 'blurring of boundaries' in the explanation without any questions. The difference is that high rises cannot be done without Engineers, but the best ones are constructed when both disciplines work at their best. At the other extreme many architects can, and do, create houses without any Engineer involvement. However the most successful houses occur when both disciplines are involved.

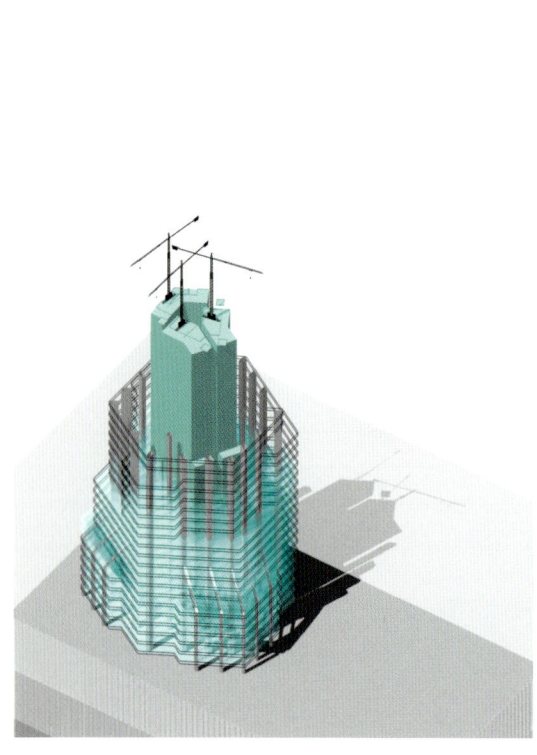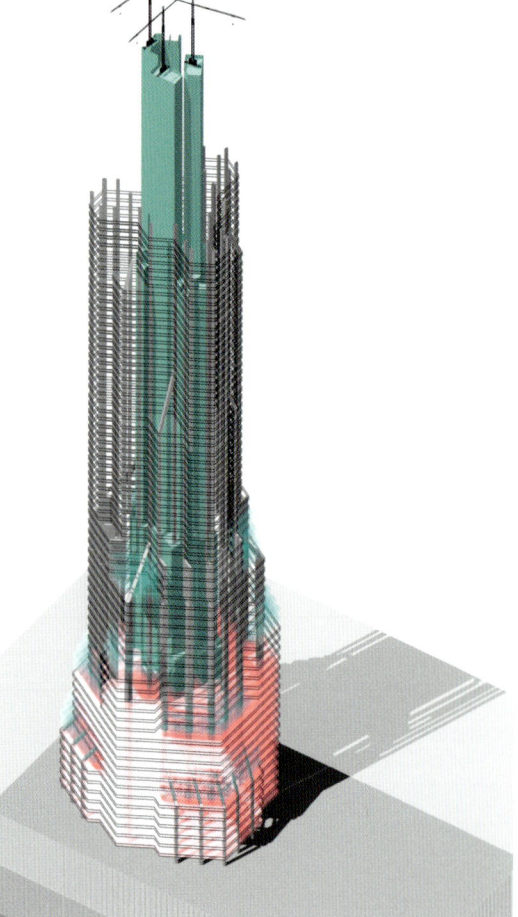

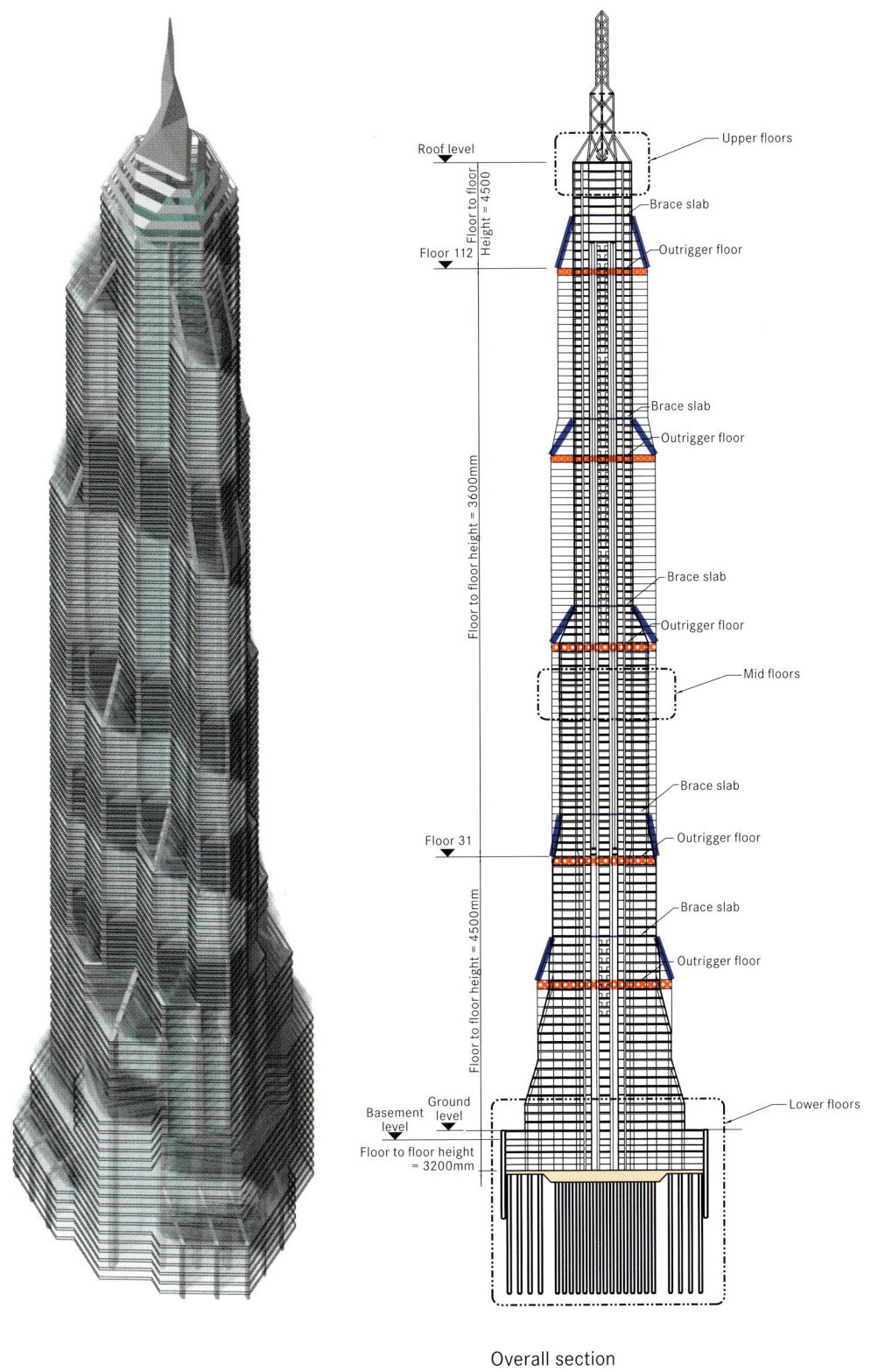

Roof level

Floor to floor
Height = 4500

Upper floors

Brace slab

Outrigger floor

Floor 112

Brace slab

Outrigger floor

Floor to floor height = 3600mm

Brace slab

Outrigger floor

Mid floors

Brace slab

Outrigger floor

Floor 31

Brace slab

Outrigger floor

Floor to floor height = 4500mm

Basement
level

Ground
level

Lower floors

Floor to floor height
= 3200mm

Overall section

UNLEARNING

Sometimes unlearning is just as important in developing new solutions as the traditional learning process. On many occasions, the office has had to dismiss past precedents in order to come up with new solutions. This happens in radical situations, like the Phæno Science Centre, where there is no known precedent for the performance that is trying to be achieved. However, this unlearning also happens in smaller projects or elements, where deliberately forgetting the conventions of a problem can be the key to rethinking a problem in a new way.

We have also found that in some instances a 'tried and tested' method or system is heavily promoted and not questioned. The productive power and speed of technologies today is such that 'goals' and 'evaluation criteria' should always be re-examined each time one works on a new project. Asking new questions and finding new answers results in new knowledge. This benefits disciplines that are motivated to create and innovate. The desire to unlearn also comes from observing that the gap between the intellectual end of the spectrum (architecture, engineering, design) and the less intellectual end (constructors, fabricators) is widening and needs to be closed.

Design thinking with Adams Kara Taylor

Michael Speaks

Researcher and editor, dean of the University of Kentucky College of Design and director of Big Soft Orange

As the limits of what can and what cannot be built are daily pushed to new extremes by an insatiable, market driven appetite for more and ever more complex and formally exotic buildings, the architect and the structural engineer have become a close, if not inextricably linked, pair. Cutting edge architects, in particular, seem incapable of building the orchidaceous shapes and forms that issue from their computers without a cutting edge structural engineer. And yet, however close, each has a professionally, legally and culturally designated role to play in the making of buildings. The architect is the creator who makes designs and the engineer the analyst and tester who optimizes and makes buildable those designs. That, at least, is the conventional understanding of their relationship. There are, of course, many notable historical and contemporary exceptions to this convention—engineers like Brunel, Eiffel, Maillart, and Nervi, who stand with one foot on either side of the line that separates the design from its optimization. But such examples of boundary crossing, or straddling, as it were, affirm rather than contradict the line itself.

Among the most compelling attempts to blur if not altogether call into question this line, occurs in Peter Rice's posthumously published book, An Engineer Imagines, where he discusses the creative role engineers play in the design of buildings. Rice begins "The Role of the Engineer," perhaps the most enlightening chapter in the book, with the following observation. "I am an engineer. Often people will call me 'architect engineer' as a compliment. It is meant to signify a quality of engineer who is more imaginative and design-orientated than a normal engineer... If people find an engineer making original designs which only an engineer can make, they feel the need to grant him or her a higher accolade, hence 'architect engineer.'" Rice goes on to point out that the engineer's work is just as imaginative as that of the architect, the principal difference between them being that the architect works more subjectively "to create" while the engineer works more objectively "to invent." Rice believes that both the architect and the engineer are designers, though each has a different design process. Architects approach a design problem from the standpoint of their own personal design signature. They are often hired, Rice notes, to solve a design problem precisely because the way they will respond and even the style of the response is known beforehand. After all, it is the solution, recognized before and after the commission in the architect's signature style, which the client seeks in selecting an architect. The engineer, on the other hand, approaches design problems as if the solution were not known in advance and therefore needs to be produced. If successful, the architect designs a "new" solution that is anticipated in the design problem given, while for the engineer, success depends entirely on designing an innovative and therefore unanticipated solution.

Rice's account of engineering innovation is remarkably similar to what design leaders like David Kelly of IDEO call "design thinking," a form of design prototyping that follows a classic distinction made by business thinker Peter Drucker between problem solving, which answers without questioning the problem given, and therefore adds nothing new, and innovation, which interrogates and reforms the problem given and adds value by creating new knowledge and new products not anticipated in the problem. Problem solving shapes the known while innovation coaxes into existence the unknown. Design thinking is a "thinking by doing" in which plausible solutions are prototyped, interrogated and redesigned. Prototypes, which IDEO call "the shorthand of innovation," are not, however, variations of a projected final design—they are not guesses extrapolated from the designer's perfect idea about what the final design might be—but are instead "what ifs" that the designer uses to drive the innovation process itself. The designer uses the prototype to "think through" as many factors as necessary—material, cost, fabrication, etc.—and adjust the design accordingly. Not only are the assumptions of the problem given transformed—opening the way for innovations—but also with each prototype new design intelligence is generated that can be shared and discussed among teams of designers whose additional input further enhances the innovation process. In such a design driven, innovation economy, where collaboration and prototyping have become more important than the designer genius and his one-of-a-kind design objects, it is no wonder that the engineer's unique design contribution is finally being recognized. Indeed, as the computer and computational thinking have come to dominate architecture and engineering, analysis and testing—the very kind of "design thinking" the engineer has traditionally been tasked with—have become increasingly important engines of architectural innovation. As a result, cutting edge architects are beginning to understand the significance of innovation for their designs and for their design practices and, with cutting edge engineers, they are entering into new collaborations that call into question, once and for all, the line between the design and its engineering.

Among the leaders forging these collaborations are London based Adams Kara Taylor, a growing consulting structural and civil engineering practice that has made a name for itself by focusing resolutely on innovation—of engineering products, processes and services. They are also among a handful of engineering practices that are redefining the role of the engineer and proving, along the way, that design, creativity and innovation belong as much to engineering as to architecture. From the founding of the firm, AKT made a concerted and strategic effort to seek out and work with architects and developers that were on the design cutting edge, even if it meant working on projects that were often fiscally disadvantageous and sometimes, due to the complexity of the engineering challenge itself, so perilous that failure would likely have threatened the very solvency of the firm. AKT considered this kind of risk-taking a worthwhile investment that would pay returns in future opportunities to work with other "difficult" architects on projects that would give them the chance to innovate. Ten years later and now 100 plus, AKT have worked with some of the most important architects in the world on some of their most challenging buildings, from Zaha Hadid Architects,

Alsop Architects, and Norman Foster to rising stars including Foreign Office Architects, Future Sytems, Thomas Heatherwick, and BIG. This past year Hanif Kara, one of the founding directors, was selected as a member of the prestigious Design For London Advisory Group, the same year AKT was chosen as the only engineering firm to exhibit in the British Council's "Best of British Design" show, adding further proof, as if any were necessary, that AKT have become one of the most important new design engineering firms in London, which means one of the best in the world.

Joop Paul, Director of Arup in the Netherlands, has observed that engineers are today not needed on most architectural projects, only on those that require real innovation. In those situations, he says, the architect and the engineer together embrace the unknown and try to wrest from it something completely new and unforeseen: "They chart new waters—in a design team—without knowing exactly where they'll end up." It was in this spirit that AKT entered and won with Zaha Hadid Architects the competition to design the Wolfsburg Science Center in Wolfsburg Germany, built, ultimately, in 2006. As is becoming more common in such collaborations, AKT was involved with the design from the beginning. Indeed, given the limits that the use of a concrete exoskeleton placed on future design changes—once the design was fixed only variations in the thickness of the shell could be made—this project, in particular, required an extraordinary amount of front end design work. Hanif Kara, however, is quick to point out that in this project and in other such collaborations the engineer's job is always to support and materialize the architect's design concept. "Its the architect being supported by the engineer; not the architect becoming an engineer or the engineer becoming an architect. In Wolfsburg it was their concept and we made it happen." What is required, however, in order for the collaboration to work, is that the architect and the engineer have an understanding, even empathy for, what the other does and is faced with. There is, Kara says, a naiveté required that enables each to take on the enormous risks necessary to design such a complex and structurally challenging building.

In Wolfsburg, technology, specifically the sharing of digital information, made collaboration easier. Both ZHA and AKT used Maya, Form Z and other software—for the architects these were design tools and the engineers testing tools. Yet, as Kara points out, none of these were able to analyze structure and materials and so it was necessary for AKT to write connecting software that linked the digital designs of ZHA with their own analysis software. Even so, the building presented enormous analytical challenges. As Kara says, "When we won the competition there was no single tool, no software that could analyze the building as a single element because of its enormous scale." This ultimately led AKT to develop a strategy to customize a commercially available software package for finite element analysis, Sofistik, that otherwise was unable to analyze a building the scale and complexity of Wolfsburg. After nine months of close collaboration with Sofistik, AKT were ultimately able to enhance the package and make the required analysis of the more than 40,000 elements of building structure.

The collaborative efforts of AKT and Sofistik to innovate the software package insured the integrity of ZHA's design concept, though this innovation and many others cannot be directly observed in the final result—not in the stunning concrete table that seems to float on a cushion of light; not in the support cones on which the table and roof structure appear precariously balanced; not in the continuous interior landscape that magically dissolves structure, building and program into a continuous though variable experiential terrain. With ZHA's German affiliate, along with the building contractors and concrete experts in Germany, AKT also developed an innovative use of self-compacting concrete to pour the support cones. Self-compacting concrete had been used in buildings before but never in such a megastructure. Such innovations and the design thinking required to produce them are not so easily seen in the building but are crucial to its design and construction. Analysis and testing enable the necessary give-and-take between architect and engineer that allows them to transform and improve the design over and again without compromising the architect's original design concept. In this way design becomes what Joop Paul calls "a game of ping-pong between architect and engineer," where the engineer's testing and findings become "input to look afresh at the design." When AKT were finally able to analyze and test the entire structure, their input allowed ZHA to reduce the amount of concrete used, which, along with the innovative use of CNC milling and rapid prototyping in polystyrene and timber to create custom formwork, reduced the use of concrete to such a degree that none of the supporting concrete structure was more than 30 centimeters thick, further enhancing the porosity, lightness and formal elegance sought in the design concept. And the same kind of design thinking dramatically affected the clear span steel roof structure. Breaking with convention, AKT were able to model, analyze and test a structural catalogue system of variable, mass customized parts that by conventional means of calculation—by tonnage—appeared unfeasible and uneconomical but when construction and welding costs were taken into consideration, proved cheaper, easier to construct, more structurally sound and, a better expression of ZHA's design concept.

The product of the engineer's design thinking, then, is not so much the design as it is the design knowledge or intelligence that comes from analysis and testing. Working in collaboration with cutting edge architects like ZHA, Alsop, and others, has certainly helped to make AKT one of the most important design led engineering practices in the world. But it has also greatly enhanced and expanded their engineering design intelligence. In projects that require seemingly constant material and structural innovation, such as the Wolfsburg Science Center, not only does analysis and testing begin much earlier, but also given the complexity of the project and the design permutations and variety of options that need to be considered, analysis and testing speed is at a premium. Being able to repurpose successful innovations that occur in such projects as well as the intelligence accumulated in the hundreds of thousands of analyses required to produce them, offers real advantages in speed and in expertise. Realizing this, AKT have taken the natural accumulation of intelligence that occurs in collaborations one step further and has created a team within their office whose principal objective is to facilitate collaboration with architects and to generate, through original research, new intelligence that can be used in collaborations and can, in some cases, be transformed into new stand alone products. The team, which is physically spread throughout the office and only comes together to work on projects, is called "P.art," and consists of architects, engineers, computer programmers and graphic designers. Because they are conversant in the tools and techniques of both architects and engineers, they are able to help engineers at AKT better understand the architect's intentions and ambitions as well as translate the work of AKT engineers for the architect.

P.art occupies what founding partner Albert Taylor calls the zone—the design space between architect and engineer where innovation is most likely to occur. And it his job, along with Kara's, to manage this zone, pulling and pushing the line it draws between architect and engineer without crossing into unproductive territory. Like Kara, Taylor believes very strongly that the engineer and architect play distinct roles in the design of buildings and structures and that the engineer should never try to assume the role of the architect. In fact, he argues that their differences create the necessary tension that makes innovation possible. While digital information sharing allows the architect and the engineer to look in, as it were, on one another's work, it is important that each does so to gain intelligence from another perspective, as Rice might say, because it is there that one will discover innovation. Taylor says that because P.art is made up of team members from a number of different disciplines each of whom often comes from a different part of the world, it internalizes and multiplies those differences and perspectives. But P.art also provides a medium within AKT to channel and transform those differences into design innovation. Some of those now include stand-alone products such as the furniture pieces P.art designed with Future Systems, as well as a bridge they designed with scripting and parametric modeling. P.art has also begun using scripting to create bespoke design-analysis packages with built in structural constraints and rules that allow architects to experiment with and think through design ideas and problems, creating, in the process, little bits of design intelligence. As Ruben Brambleby, leader of the P.art team says, they wanted the architect to be able to play around with their ideas but know, at the same time, that no matter how far they might stray, the rules and parameters would pull the design back within the realm of the possible.

Taylor, like Kara is also interested in the zone of innovation and collaboration between engineering and fabrication. Because the engineer often works closer than the architect with those who make things, he acquires an intimate knowledge of the processes, techniques and materials used by everyone from the craftsman who makes things by hand to the operator of the 3D printer. Kara was once a welder and like Taylor has deep personal knowledge of and affinity for making. Because AKT is one of the few engineering offices in London, and one of the very few design led firms in the world, to have all of their drawings in a digital, 3D format, they are able to move very quickly from physical to digital to analytical models and now even to fabricate directly from them. But this does not mean that the design expertise and intelligence created with the physical model is any less important than the digital 3D model. Each has a design intelligence quotient equal to the task it is required to complete. Whether hand crafted or digital, each model, prototype or simulation is nothing more than a means to "think by doing." One can learn from even the crudest means of design thinking. That is why it is important, Taylor says, to put in the same effort working with the fabricator, even if a single craftsman who cuts templates from a one-to-one-printed Maya file with a jigsaw, as when working with famous architects. Given the unequal speed of technological adaptation, it is likely that mash-ups using cutting edge and seemingly outdated technology will become more not less common. It is crucial, then, as Taylor says, that AKT move in and between these various design perspectives—from Future Systems to the jigsaw craftsman—making use of the special design intelligence produced in collaboration with each. That, in fact, is how AKT have managed their very unique zone of innovation for more than ten years now, adding, with each project and each collaboration, to their already prodigious storehouse of design knowledge and expertise. Guided by such intelligence, AKT are sure over the next ten years to expand their zone of innovation and in so doing help to make engineering the kind of creative design practice that Peter Rice always knew it could and someday would be.

General information of the Phæno Science Centre
see chapter COMPLEXITY, case CONES, page 82

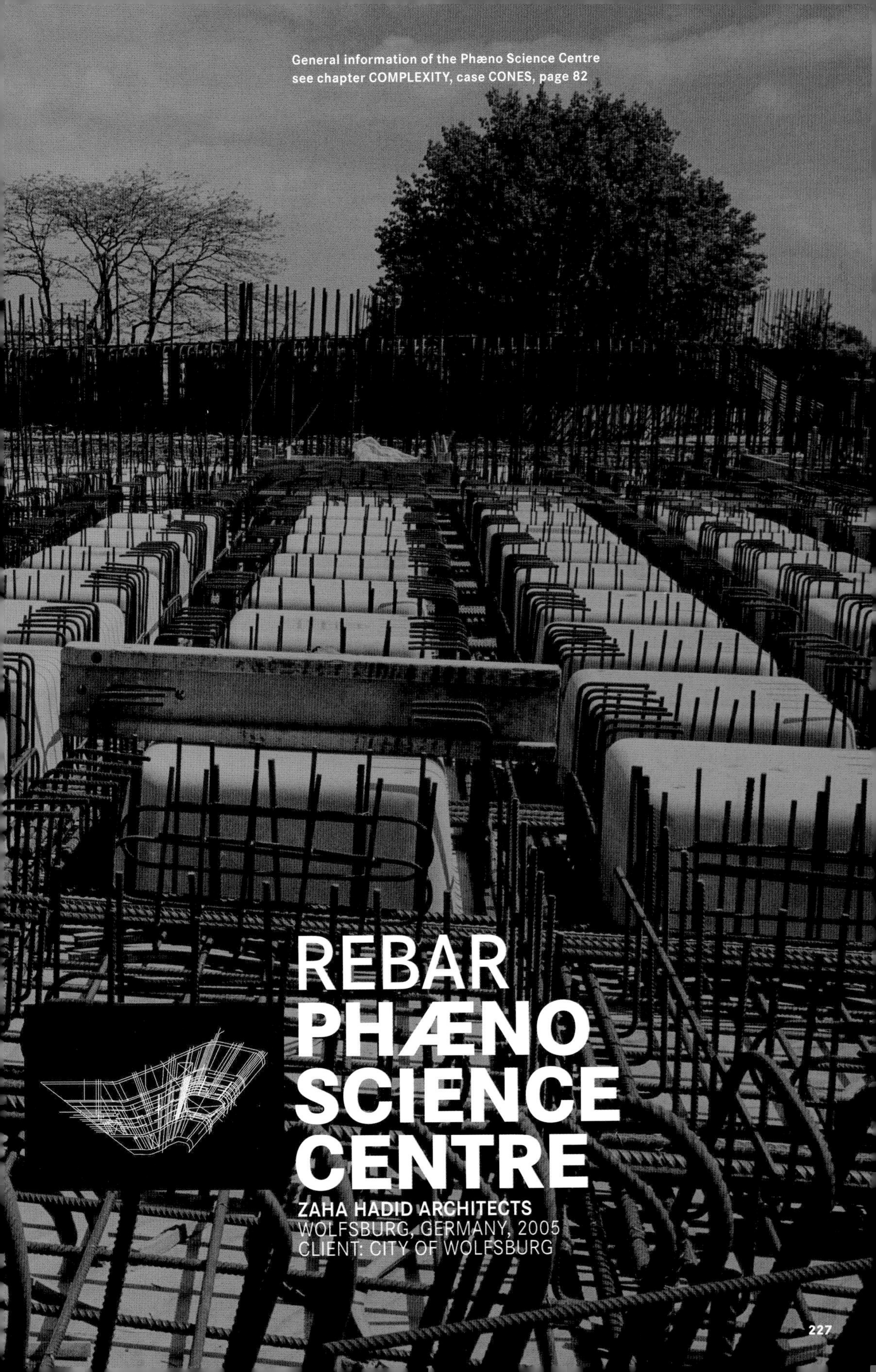

REBAR
PHÆNO SCIENCE CENTRE

ZAHA HADID ARCHITECTS
WOLFSBURG, GERMANY, 2005
CLIENT: CITY OF WOLFSBURG

Given the geometrical complexity of the Phæno Centre, how was the reiforcement determined?

From a design viewpoint, the whole project was about communicating intensively in 3D. The teams were totally focused on how to unfold and shape the building to reassemble it, the most primitive component of this process being the reinforcement bar.

Take for instance the sloping cone walls. These were developed using tangents and radii from which timber formwork was produced. The sloping and curving of this surface had vertical and horizontal bars like a conventional wall, so the reinforcement drawing is a flattened version of a conventional wall. This was disappointing to the architects, since the complex analysis they had seen in our work implied all sorts of diagonal patterns in stresses and differentiation in stress levels, and they had expected this to be reflected in the reinforcement bars. We are not at that stage technologically as it would mean replacing the bars with fibers that could support large tensions. Whilst fiber reinforced concretes are on the market, they are mostly used for low stressed elements such as ground slabs.

Slicing and cutting the building into parts, which could then be reinforced in a conventional way, became one of the biggest challenges on site. As a result of this, we were unable to 'prefabricate' the reinforcement and it had to be assembled one bar at a time, so the three dimensional process was broken into two dimensions, and ultimately a single line representing a reinforcement bar.

Any discussion on the construction of the Phæno Centre has to consider the 'macro and micro' scales to make sense. From the 'macro scale' the building was analysed as a monolithic total system. Using a finite element packaged called Sofistik, built out of a model with more than 10,000 elements. It is only due to the recent advances in digital technologies that this could be permitted. But at the 'micro level' the task was to produce a reinforcement bar drawing.

Every single reinforcement bar had to be classified, tagged, prebent and laid out on site. This is a labour intensive process, particularly when we move into complex forms in concrete, where the majority of the construction cannot be pre-cast off site in factory conditions. Pre-fabricated cages are now common for conventional buildings and indeed we used those for the main waffle floor. However as the driving forces in these are assembly, size and crane size sometimes you end up with more reinforcement than is needed.

A further complication is that the construction takes place in a 'bottom up' sequence and each element has to be temporarily supported until it can bear its self-weight. This has to be reflected in the stages of analysis. The building acts as a 'whole' only when the last slab is completed.

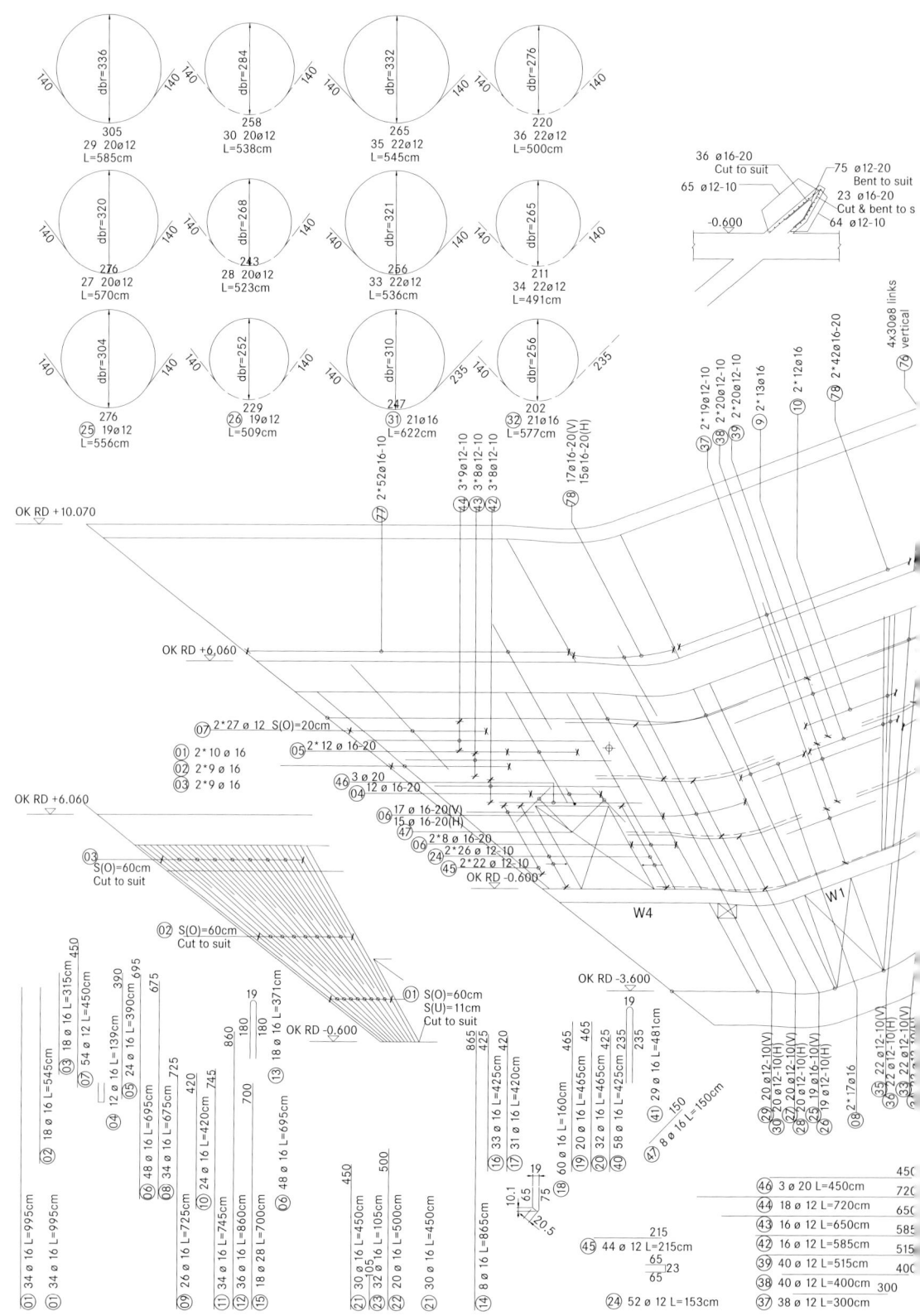

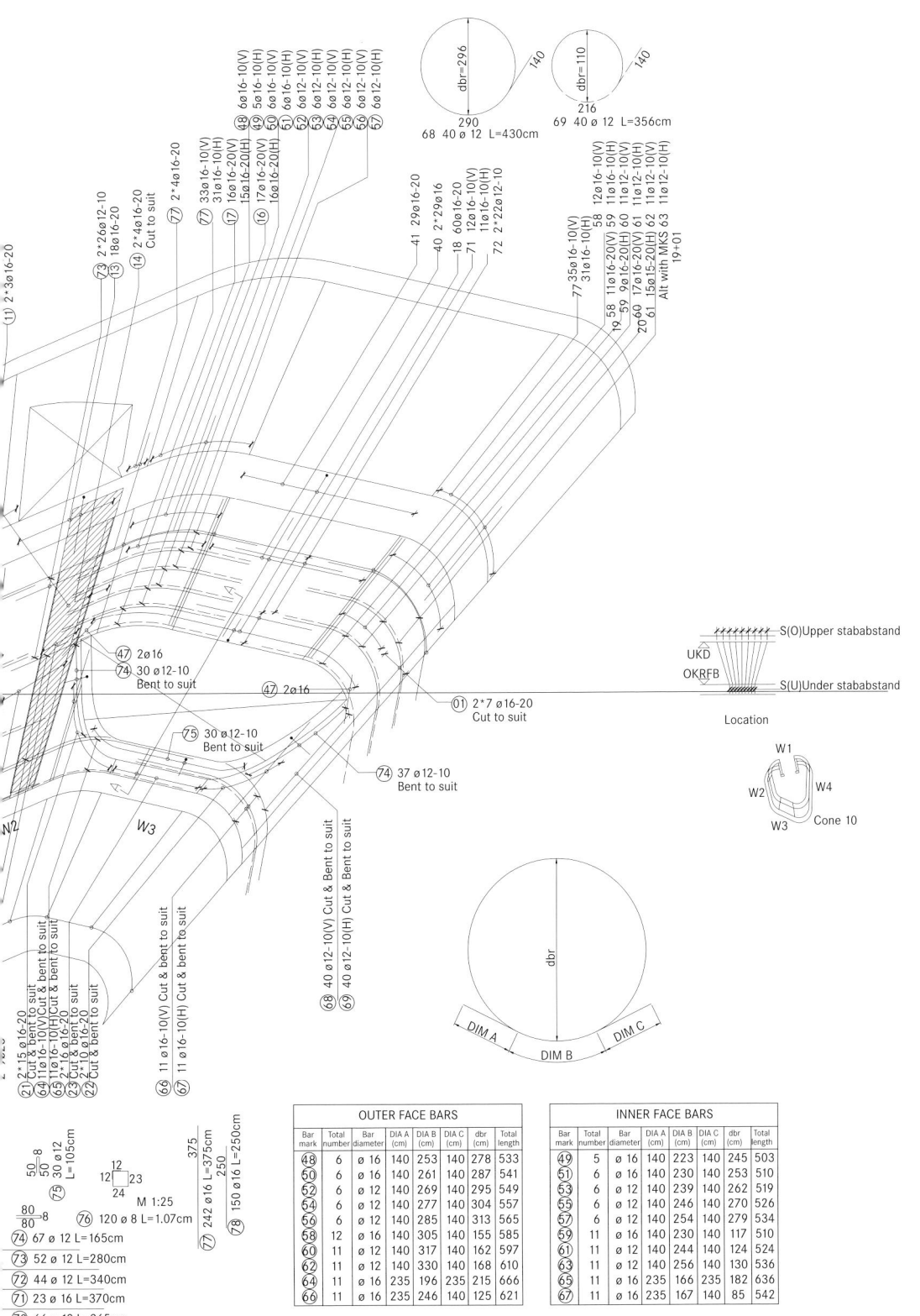

OUTER FACE BARS							
Bar mark	Total number	Bar diameter	DIA A (cm)	DIA B (cm)	DIA C (cm)	dbr (cm)	Total length
48	6	ø 16	140	253	140	278	533
50	6	ø 16	140	261	140	287	541
52	6	ø 12	140	269	140	295	549
54	6	ø 12	140	277	140	304	557
56	6	ø 12	140	285	140	313	565
58	12	ø 16	140	305	140	155	585
60	11	ø 12	140	317	140	162	597
62	11	ø 12	140	330	140	168	610
64	11	ø 16	235	196	235	215	666
66	11	ø 16	235	246	140	125	621

INNER FACE BARS							
Bar mark	Total number	Bar diameter	DIA A (cm)	DIA B (cm)	DIA C (cm)	dbr (cm)	Total length
49	5	ø 16	140	223	140	245	503
51	6	ø 16	140	230	140	253	510
53	6	ø 12	140	239	140	262	519
55	6	ø 12	140	246	140	270	526
57	6	ø 12	140	254	140	279	534
59	11	ø 16	140	230	140	117	510
61	11	ø 12	140	244	140	124	524
63	11	ø 12	140	256	140	130	536
65	11	ø 16	235	166	235	182	636
67	11	ø 16	235	167	140	85	542

So if we focus on the 'truncated cones', each is formed by flat walls with radiused corners. Each one has a unique geometry, and the wall inclines from about 35° to 90°. The walls are then also further complicated by the need for openings at ground level to allow doors, entrances, etc.

The architects were very specific about the formwork to achieve the desired finish so they had to produce formwork drawings (timber planks in this case) which in turn had to relate to the reinforcement drawing we produced.

So again we were going from producing 3 dimensional drawings to collapsing this through 2D into plank shutters. The planks were butt jointed vertically with a maximum width of 305mm and cut conically (tapered) where the wall was conical.

As we detailed the reinforcement, it became clear that to get the correct fair faced finish and deal with congestion of reinforcement, self-compacting concrete had to be used.

This was poured in heights of up to 7m, thus exerting large hydraustatic pressures on the shutters as the concrete hardens.

Was the reinforcement and formwork the same for the coffered slab?

The transition between the cone walls and the ceiling resulted in double curved walls. This had to be accurately built with small radii; we had to use moulded GRP elements. The waffle slab was made out of CNC cut polystyrene blocks.

Feilden Clegg Bradley's design for student housing in East London reuses a former goods yard located between a railway and a canal.

A series of E-shaped, four-storey buildings with a taller block on stilts, the project contains 1,000 units. AKT's engineering proposal had to take into account exisiting site contamination and a green roof. While steel, concrete, or load-bearing masonry would work as a structural solution; the optimum solution is a balance between environmental sustainability and ease of construction. Concrete tunnel form construction was ultimately chosen:

In this technique, concrete is poured around moulds, which are removed and reused when the concrete sets.

At 3.15 x 6.3m, the basic unit module is small enough that is does not to require extra stiffening. The high amount of repetition allowed for the use of very high quality moulds tht would provide a high standard of finish. Quickly built, the structure is strong enough to support a "green" roof.

CONCRETE FRAME, POPOUTS
QUEEN MARY STUDENT VILLAGE

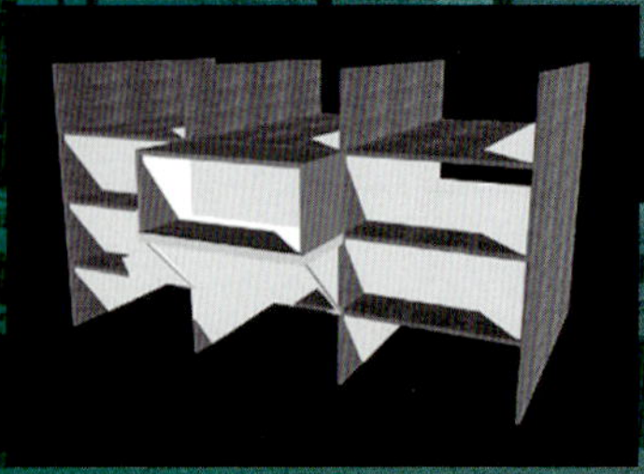

FEILDEN CLEGG BRADLEY STUDIOS
LONDON, 2007
CLIENT: QUEEN MARY COLLEGE, UNIVERSITY OF LONDON

Why and how do you think this project contributes to the work?

A journalist once described this project as one where 'bling meets budget' which makes a good starting point to discuss the role of this project. In an age where complexity and complex forms are evermore prevalent, the simplicity of projects such as this one can very often be easily overlooked. In order to exploit and generate ideas and creativity we have to fall back to historical paradigms, and the 'housing' type gives just such an opportunity. Dormitories and student housing at this time were less regulated than standard 'housing' due to the transient nature of the occupants. However, this trend is now changing. For instance, the acoustic requirements between spaces are not as stringent as housing and it falls onto the designers to explore appropriate options. Here, our strategy was aimed at the heart of the process used to design and deliver these options. The general approach is to limit the role of the designer and to allow contractors to deliver; the result often being a repetition of the designer's previous project, with little or no revisiting of specificity to the project.

How was involvement different?

The opportunity of building 992 units in 90 weeks was what stimulated us right from the outset. The approach we took was to challenge the notion of constructing the building in traditional masonry blocks. The master plan emerged from the brief and site context, but also, at this stage, a 'structural massing' which included a number of 'courtyard buildings', one eight 'storey block' and a 'pavilion block'.

The process involved examining all possible ways of constructing the frame and agreeing on a rigorous evaluation matrix. We were thus able to re-introduce 'tunnel form' construction back into the market by proving that reinforced concrete would be the right choice and that by 'modularising' and assembling blocks between 5 to 8 storeys, we could construct the building. It was a technique that was common in the UK in the 1960s but had largely been faded out. Here, as well as delivering a robust and soundproof finished product, we also promoted the idea that it would be the fastest construction method.

The 'minor' rethink on this got us away from traditional masonry and other such construction. This also released the architect to consider better finishes for the external skin and in particular the 'pop out' kitchens on the railway elevation. This has become a recognisable building as a result.

The idea of utilising an old technology and making it viable, or 'shiny', again has been the success from our perspective. The project is also used as a Case Study in the office for training engineers to 'relearn' some of the processes that will enable them to invent practical solutions and methodologies. Furthermore, we have subsequently re-used tunnel form in a number of projects. Indeed, it has expanded our involvement in the housing sector, where we have found this approach of taking existing systems and adapting them to be very successful.

Why then do you think this technology has fallen from use over the recent years?

This project is unique in the UK, but in France and Spain the system is still used a lot. The delivery of projects like this has been the domain of so-called sector specialists. The 'design and build' route dominates the sector and as costs are low, there is little or no mind set to questions. In fact, the gap between the 'subcontractors', who are the craftsmen, and the designer has become greater as major contractors are forced to simply manage the process and question little. In this case we were quite fortunate because, prior to the contractor being appointed, we had conducted a lot of work with the Belgium based tunnel form manufacturer. We were therefore able to show the main contractor a rigorous logic. He, in turn, was keen on concrete and on investing in this system.

Why do you think other contractors on this project were not keen on tunnel form?

Mostly because it appears to be an inflexible system that relies on a tight '24 hour cycle' for building a room. Therefore, instead of seeing this as an opportunity, many contractors instead looked at the risks involved. The architect had previously considered a similar idea, but really saw the opportunity in the joint early work we did together. This previous work gave him some potential to design well on a low budget, rather than see it as an obstacle.

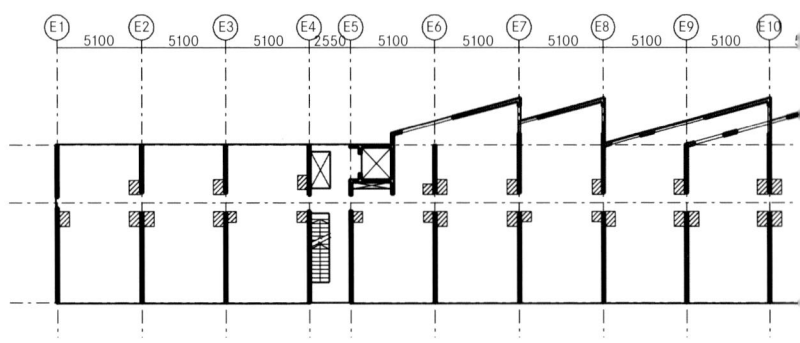

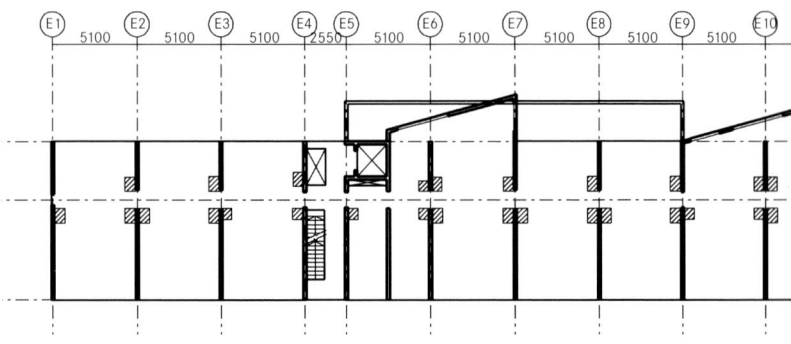

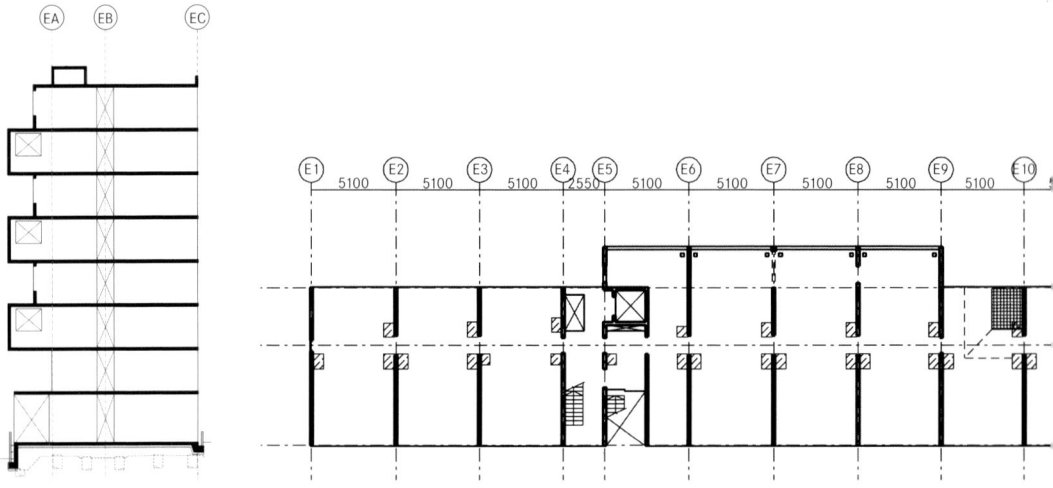

Typical plans of 8 storey block

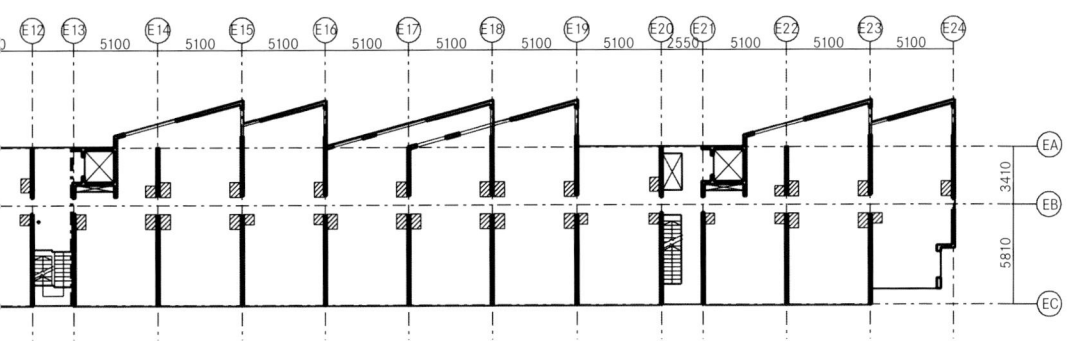

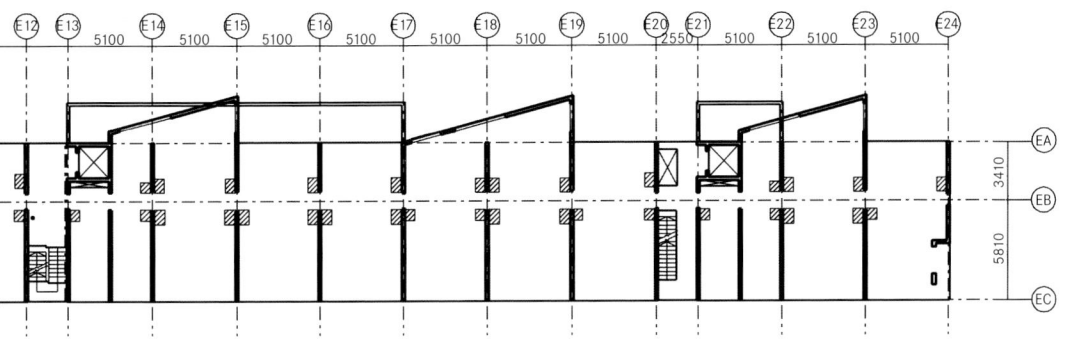

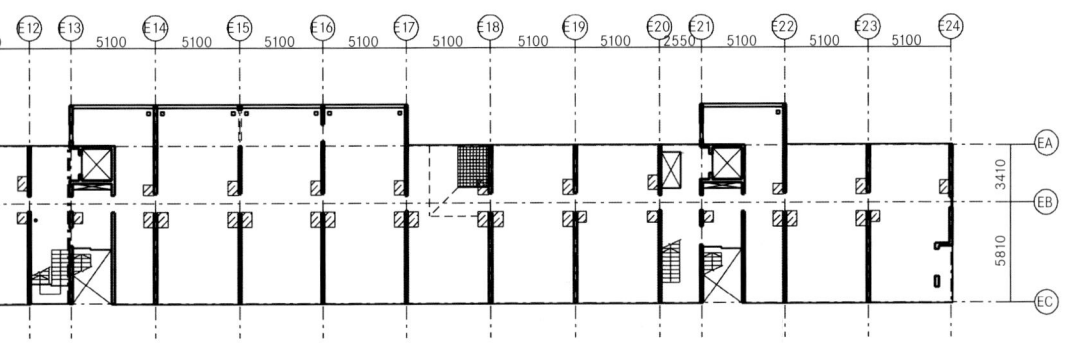

The National Trust maintains vital pieces of Britain's heritage from country houses to scenic landscapes; its headquarters, designed by Feilden Clegg Bradley, consolidates its various departments on a site located on Swindon's Great Western Railway, itself an important piece of industrial heritage. The structure is of a congruent scale and made of similar materials as the nearby sheds. A saw-tooth roof reflects an industrial typology and maximizes daylight and natural ventilation. Conscious of sustainability, this profile maximises the amount of sunlight that falls on the photovoltaic-panelled roof, providing a large percentage of the buildings electricity. The structural system is a basic steel frame, and by exposing the pre-cast concrete slabs that make up the roof, the thermal mass is used to cool the building.

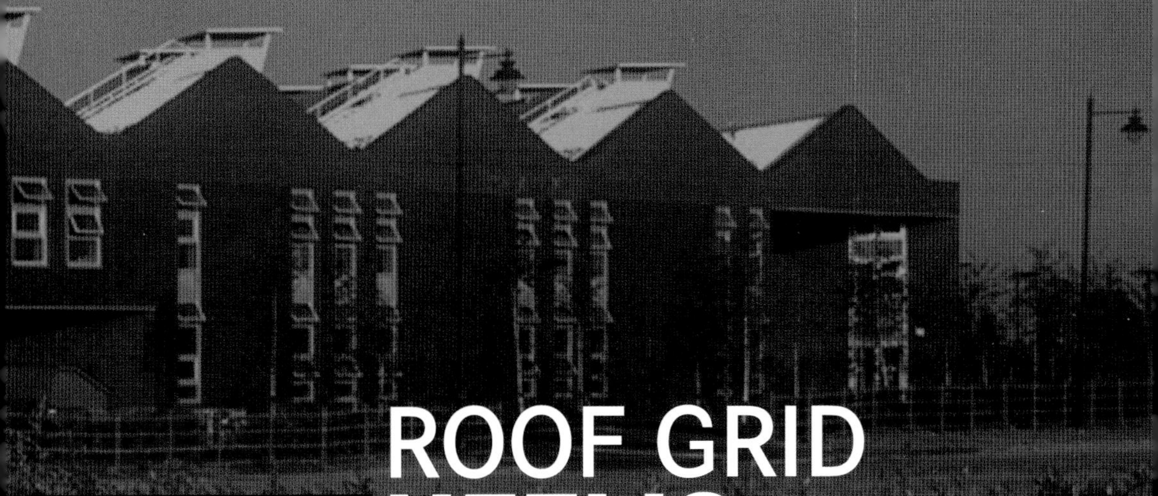

ROOF GRID
HEELIS NATIONAL TRUST

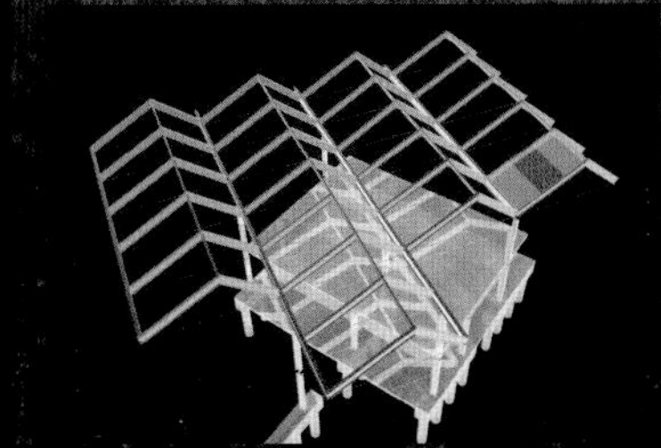

FEILDEN CLEGG BRADLEY STUDIOS
ENVIRONMENTAL ENGINEER
MAX FORDHAM
SWINDON, UK, 2006
CLIENT: NATIONAL TRUST

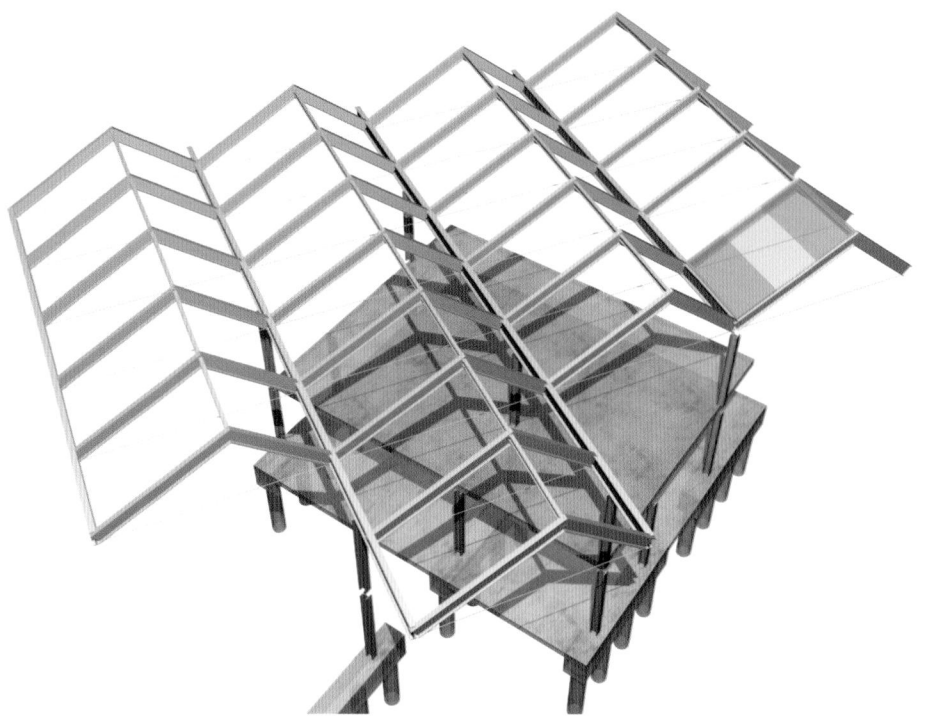

In this case the building location (Brunels Great Western Railway work) and 'shed type' imposed enormous respect and control. We had the challenge of working with the past precedence of tough industrial architecture at a time when Architects were perhaps not involved too much in this building type.

The Architects considered 'the deep plan' to be the key driver in this project, thereby making natural light and environmental control major tenets of the project. After a short interview, we were selected by the Client and Architects, perhaps due to our open-minded approach to reinventing the 'shed' with the architect. The eventual solution is a trapezoidal plan form that best responds to the site geometry, and is simultaneously adjacent to the North-South roof for solar energy and controlled natural light.

This makes the 'roof', rather than South façade, the main energy modulator.
This re-orientation of the roof gave intriguing options for diagonal grids and a number of opportunities for looking at reinforced 'concrete folded plate roofs', but the final solution relies on an orthogonal grid with a diagonal roof ; thermal mass is provided by suspending concrete tiles under the roof. The brick elevations, on three gables, incorporate brick fins and 'roof snouts' and photovoltaic panels eliminate East and West sunlight.

The project required additional sustainability measures, so re-using steel was a consideration. In addition, the variety of blue brick walls were constructed using pigmented lime mortar to minimize use of cement and facilitate recycling.
For us it became an exercise in learning and unlearning from the shed type.

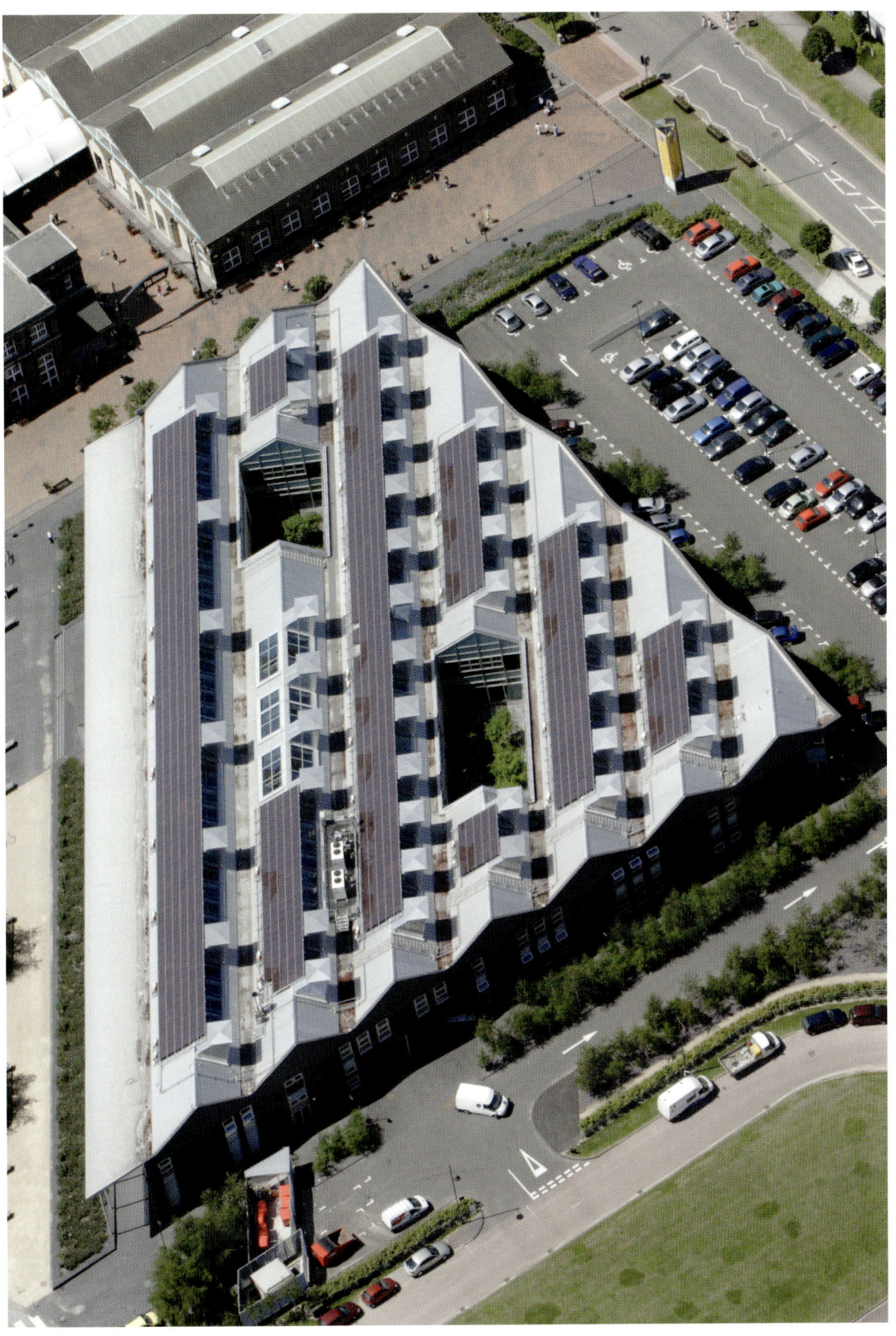

General information of the Peckham Library
see chapter PROCESS, case PODS, page 16

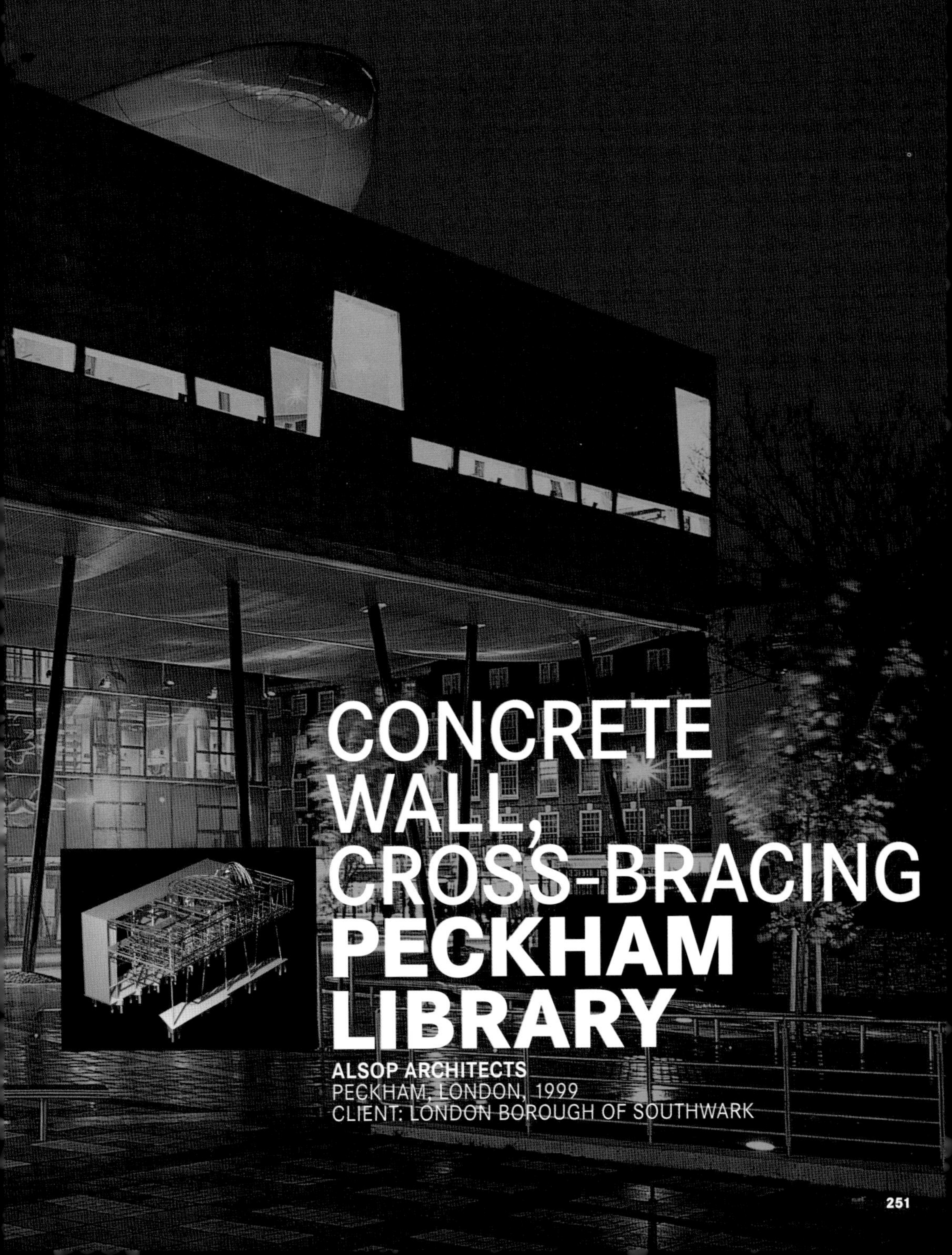

CONCRETE WALL, CROSS-BRACING
PECKHAM LIBRARY

ALSOP ARCHITECTS
PECKHAM, LONDON, 1999
CLIENT: LONDON BOROUGH OF SOUTHWARK

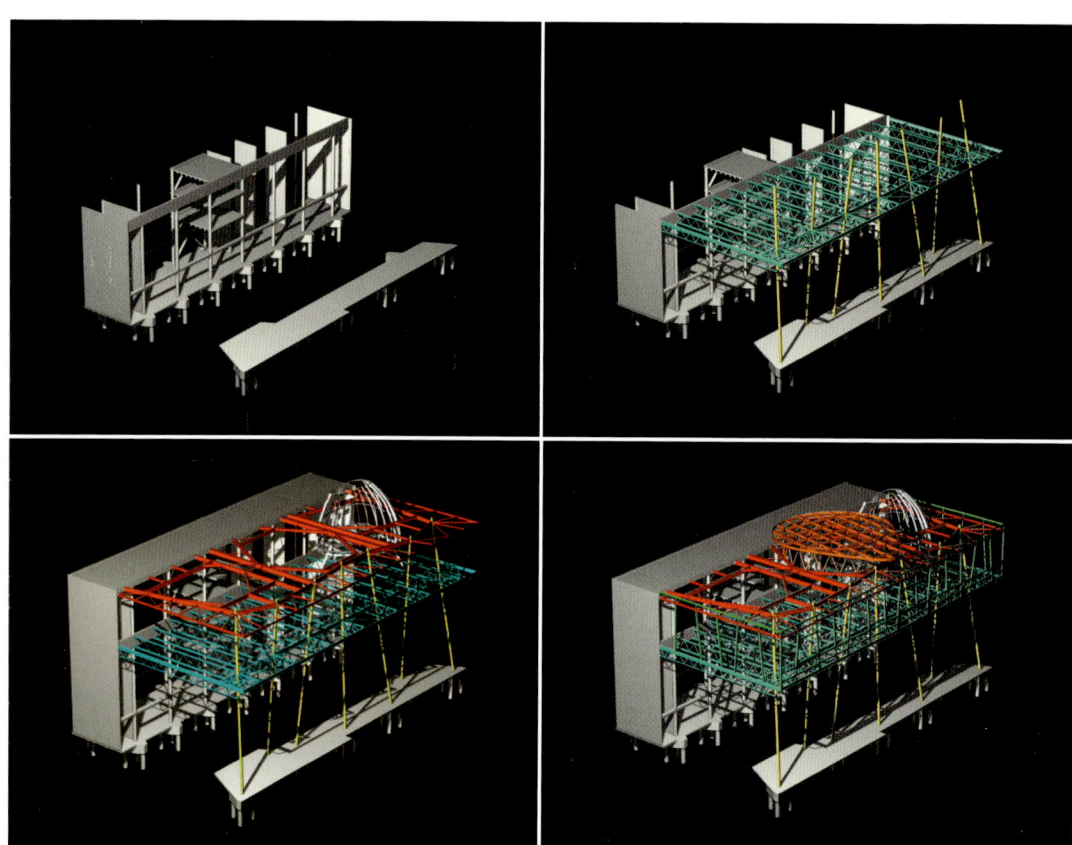

Is there an interesting story about the concrete and the concrete bracing at the back if the Peckham Library?

The project is really very clear - the vertical block acts predominantly as a large column in compression, and is short span. It was also important to expose the underside of the slab for thermal mass, so the most appropriate solution was to use the concrete construction for this section, and span the horizontal steel piece onto this.

The transverse stability of the building then required a sheer component on the rear elevation. An obvious answer would be a sheer wall, but that would not have been sympathetic to the elevation. We proposed a reinforced concrete 'k' system that could be part of the elevation, but initially had a lot of resistance from the constructor, as diagonal shaped elements on site, such as this, are not very simple to construct. This work has again helped later on, as the leaning cones of Phaeno and the 'net grid' on the Monsoon building. Therefore, these projects have allowed a continuation of this work.

The Bankside Pavilion is Zaha Hadid's winning competition entry for the Architecture Foundation's permanent home, which is near the Tate Modern. The cast-in-place concrete ribbon contains spaces for various functions, notably a gallery. Two cantilevers (one houses the gallery) project from two, twenty metre-high cores, to meet at an unsupported corner. Defining a larger distorted cuboid volume; the shape this defines is enclosed by a secondary steel glazing structure to form an atrium. AKT used finite element analysis to identify how the forces run through the structure. The process revealed places that would require extra reinforcement, others that could withstand more loads. Vertical forces are brought to the ground in two cores, which are located at the east and west ends.

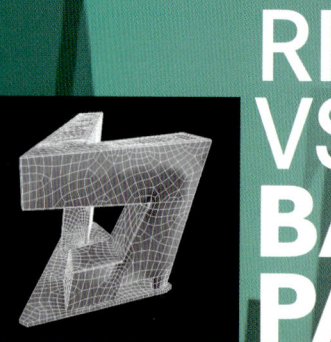

RIBBON (TUBE)
VS. BOX
BANKSIDE
PAVILION

ZAHA HADID ARCHITECTS
LONDON, 2005
CLIENT: LAND SECURITIES / ARCHITECTURAL FOUNDATION

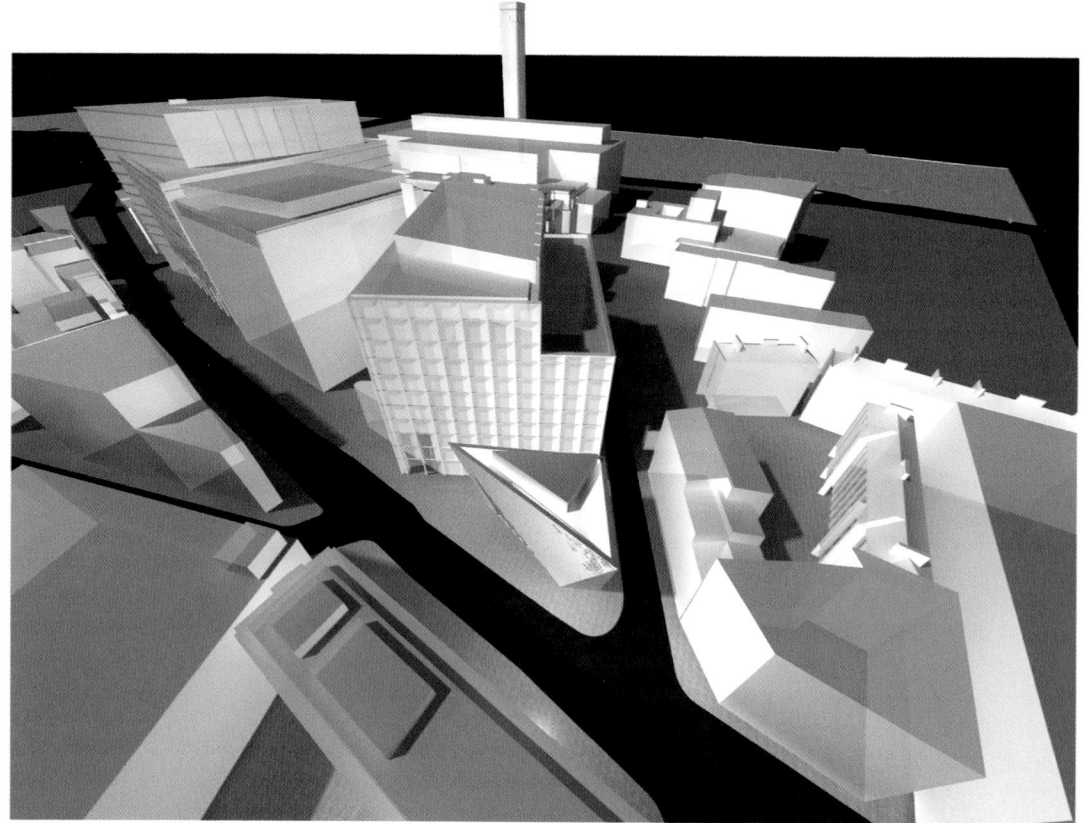

Visualisation by Zaha Hadid Architects

How did this collaboration came together?

The idea of folding a 'ribbon' had been around in Zaha's work for sometime. In fact, she reminded the jury of the private house that she had drawn, but that was yet to be built. Consequently, it was really through our collaborative work on Phaeno that the two offices became comfortable with how to develop this type of project. It must be said though, that the starting position on this was very different to that of Phæno's, as due to the triangular site, the Architect had already fixed the idea of the ribbon.

Our reading of the project is very different from Zaha's, as we read the structure as a 'folded goal post', rather than folding a ribbon. It is interesting how two different readings can survive, but this perhaps shows the shifting positions coming out of tools today, as without the right technology, one could not have been certain the cantilever would perform structurally.

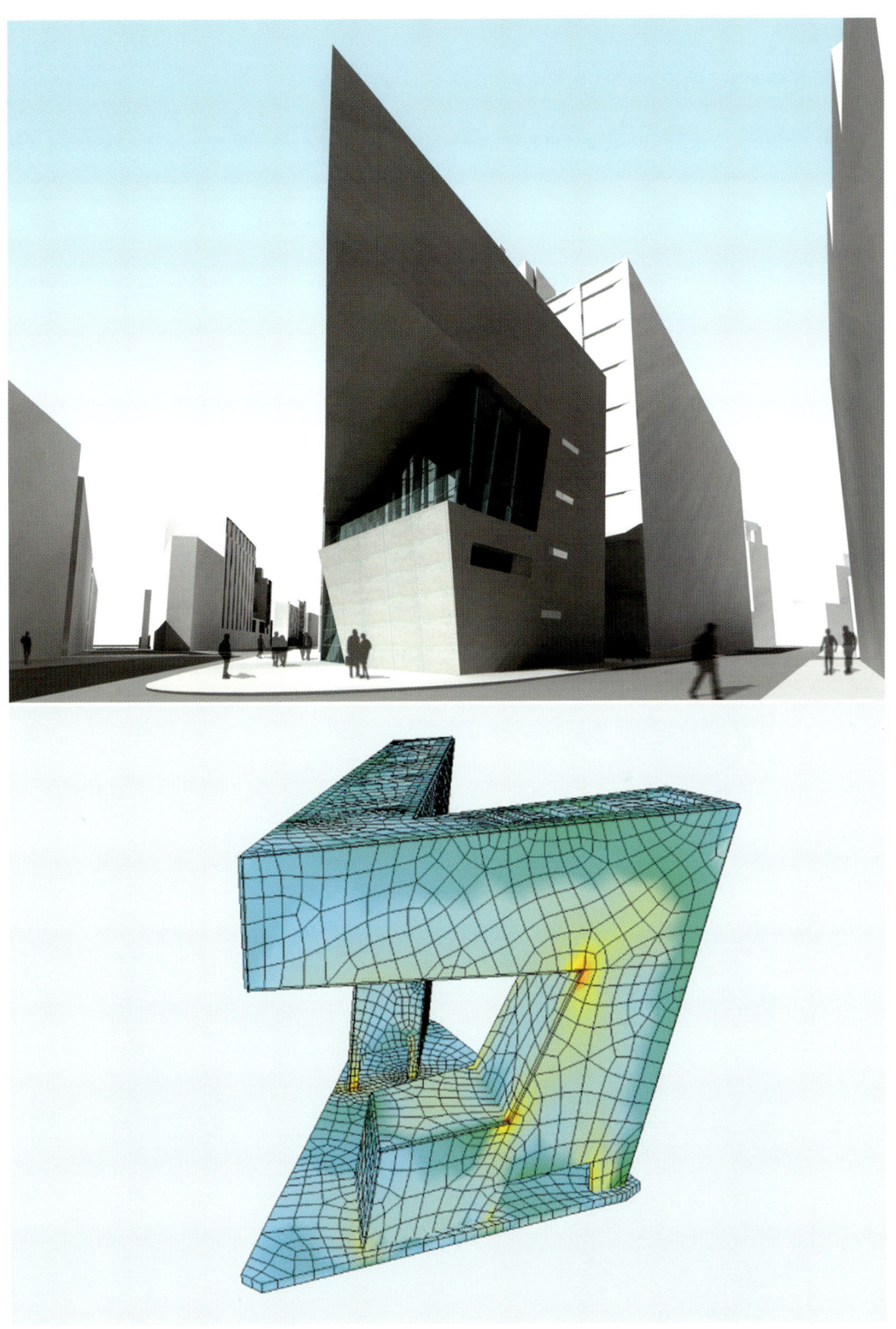

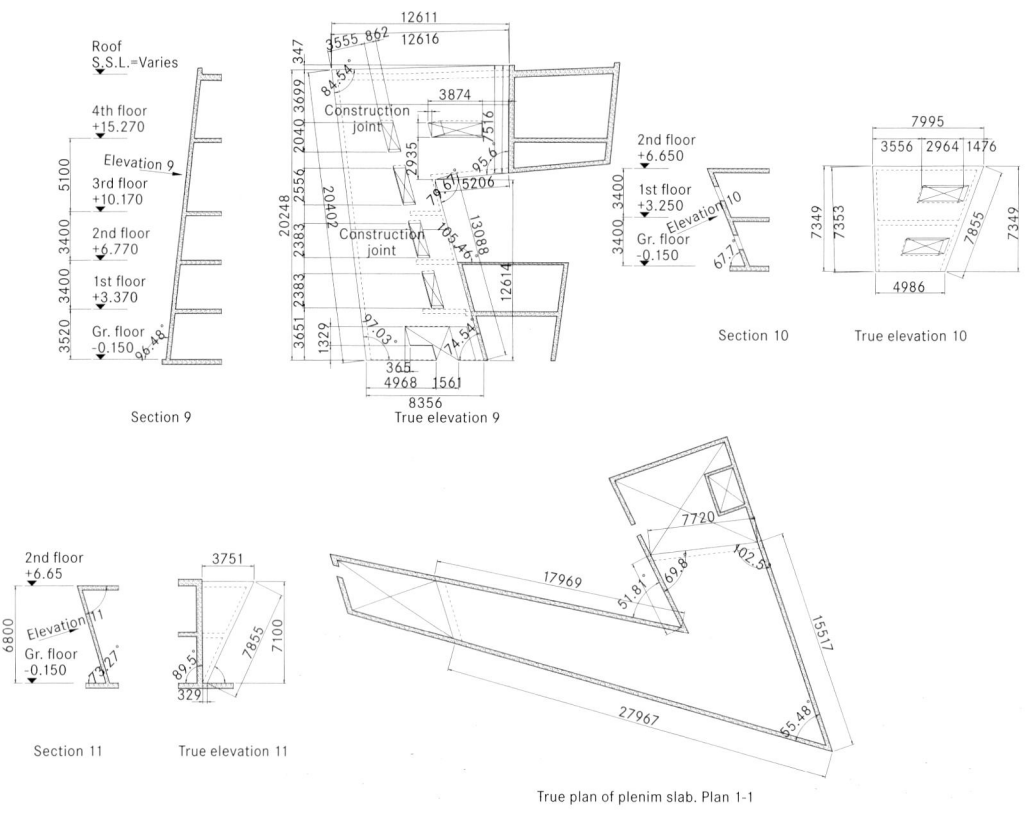

Section 9   True elevation 9   Section 10   True elevation 10

Section 11   True elevation 11

True plan of plenim slab. Plan 1-1

Why did you then choose concrete as a primary material - surely it's heavy on the cantilever?

Mostly because of the scale of the project. We recognized that the floor of the cantilever would need to be concrete anyway, as would the vertical components which act as cores. In addition, the 'ribbon' becomes a box beam and the bi-most difficult forces arrive as torsional forces onto the vertical part from the cantilever. The work on Peckham Library and Phaeno had given us confidence that constructing this would be possible.

Many comparisons have been made between this and OMA'S CCTV building - what do you think?

Technically, we are dealing with almost insignificant force magnitudes compared to CCTV. Not only is the scale of the CCTV building very different, but it also has to deal with being situated in an earthquake zone and is part of an entirely different cultural context. The Bankside project proved difficult in terms of cost and getting UK Contractors motivated, and the culture is also different in our case. Furthermore, Bankside played very much with an opaque 'ribbon', as opposed to a transparent one.

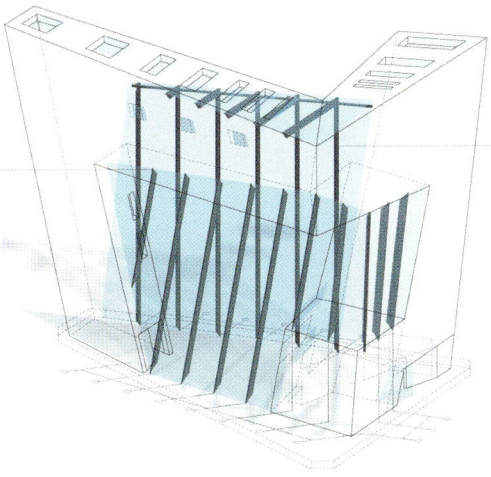

Steel I Beams at 3m centres supporting single glass panels.

Portalised I beams at 3m centres supporting single glass panels.

Façade glazing

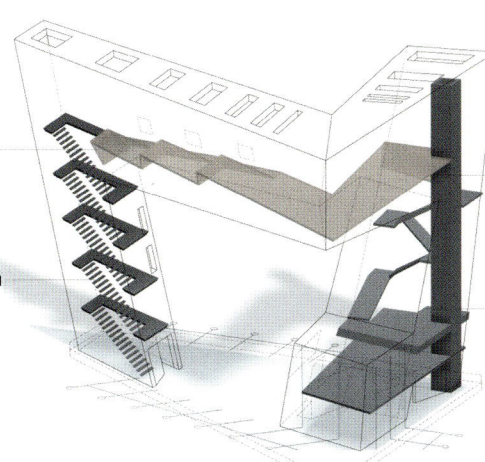

Lightweight framework forming raised floor

In situ reinforced concrete lift core

Steel staircases with in situ infill treads.

In situ concrete floor to office floor and bar terrace.

Floor plates and core

In situ reinforced concrete raft foundation

In situ reinforced concrete inclined cores.

Concrete enclosure

All visualisations of alternative scheme by Zaha Hadid Architects

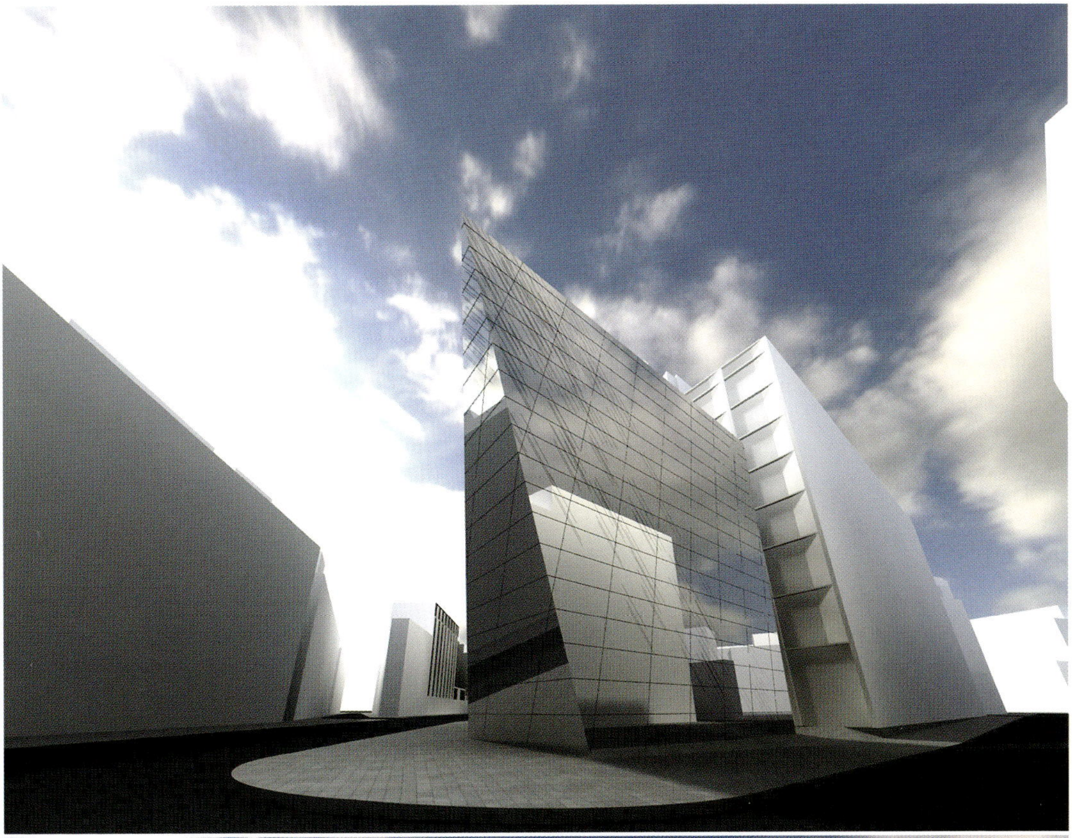

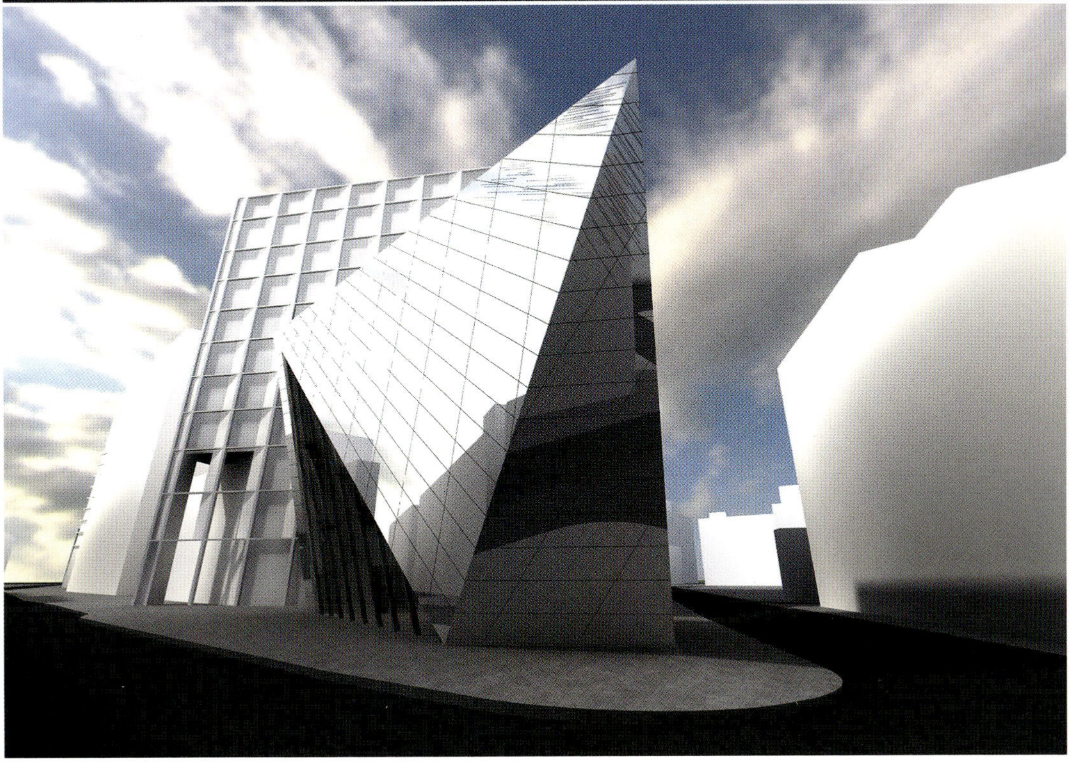

INTERSECTION

New modelling capabilities have produced new architectural potentials. Previously, we could only conceive or imagine what we can now visualise. The new capabilities of modelling tools, developed by structural engineers and other disciplines of the built environment, are making us aware of material organisations that were not perceptible before the development of sophisticated visual interfaces. Architecture and engineering, traditionally based on extensive magnitudes such as weight and length, are now visualized as intensive fields, resulting in an enormous formal shift in the generation of architectural form. Formerly, joints and junctions traditionally represented for the engineer the last visible moment prior to entering this forensic 'less' visible behaviour of forces and materials. New possibilities of registering and modelling physical organisations in time, and modelling dynamic phenomena, have opened new fields in the imagination of architects and engineers. Taking a single natural force as an example, the 'gravitational sculptor' can be shown to have evolved through the advancement of modelling behaviours and materials.

The recent intersection of disciplines, such as teaching and practicing at the same time, has begun to have affects on AKT projects. Combining the expertise of engineers, computer scientists, and architects has produced a new pattern of production in the office, where these three come together to develop new techniques and new processes for problem solving. The case of p.art (Parametric Applied Research Team), AKT's in-house laboratory testing the relationship between software, engineering knowledge and design techniques, has been the primary means for AKT to pursue the crossover between these disciplines. This highly specialised team ties together the production of knowledge from both academic and professional contexts. If engineers in the past played the role of forming or materialising ideas and designs, through the use of technological prowess, then today their greatest opportunities lie in reviving these traditions by proactively engaging in an interdisciplinary world to master other disciplines without giving way to the temptations of becoming less expert in their own discipline. It is this approach that will reshape our own discipline and produce new and revolutionary ideas.

Between engineering and architecture

Mohsen Mostafavi

Dean of Harvard University's Faculty of Design, former chairman of the Architectural Association School of Architecture and dean of Cornell University's College of Architecture, Art and Planning

In a recent set of conversations with engineers, the German magazine *Detail* asked Hanif Kara what was different about working with internationally-known architects, as opposed to other, perhaps more conventional, practices. Part of his answer sums up AKT perfectly: "The tradition in the relationships has been that the architect is the creator and the engineer the problem-solver. We subscribe to this, but operate somewhere between the two."

Hanif Kara and his partners are in many ways extremely pragmatic, both about the realities of running what is by now a sizable engineering practice in a city like London, and about the fact that – for all the discussion about architecture and engineering as a collaborative practice – the traditional relationship between architects and engineers still remains the prevalent mode of working. Architects design buildings and engineers make them stand up: in short, they solve the problems. But as Kara's response attests, AKT are not only problem-solvers but "operate somewhere between the two", between creativity and technical ingenuity. So irrespective of, and sometimes despite the client, even normative projects are considered in both creative and innovative terms.

AKT's capacity for collaboration turns them into a more pluralistic practice akin, perhaps, to the operating model Ove Arup had in mind when he founded what is now a global engineering practice. But whereas the blurring of the boundaries between architecture and engineering has led, in some quarters at least, to a blurring of roles and responsibilities as well, AKT remain sensitive to the importance of disciplinary distinctions as the basis for collaboration. They are keen to participate in the creative process but do not see their role, despite its impact on the design, as one that requires them to think of themselves as architects.

For a fast-growing office like AKT, the obvious challenges are the issues of quality and consistency. More precisely, how can they sustain and transmit the quality of knowledge and service that they were able to provide when relatively small, and furthermore keep that quality consistent across all their projects? How can they ensure that the practice's specific ethos and manner of collaboration is adhered to throughout such an expansion? Even as a very large practice, Arup maintains a model of smaller teams with their own leadership and to a large extent distinct areas of expertise. By contrast, AKT have so far resisted the formation of such divisions and instead appear to have pursued a model that in some respects parallels the academy. Not only are new engineers trained in AKT's specific ways of working to establish a consistency of approach, but the organisational structure of the office is conceived explicitly to develop new forms of engineering research.

There is of course an established tradition of cooperation in London between some of the schools of architecture and key engineering practices such as Arup and Buro Happold. In the past, this was always something of a one-way arrangement, with engineers going to schools of architecture and little traffic in the other direction. But it was quite clear from the beginning, when I first invited Hanif Kara to co-teach a design studio at the Architectural Association during the latter part of the 1990s, that his involvement would lead to more than these earlier models of academic cooperation. Most engineers, when given a chance to collaborate in architecture studios, tend to see their role in terms of supporting the work of the students. They make sure that their projects are possible to construct, without necessarily questioning the architectural ideas. Kara's engagement in the studio differs, in that his enthusiastic approach and enquiring mind generate not only technical solutions but invariably also an organisational and spatial logic which tends to have a significant bearing on the architecture as well. At the same time, he has always been clear about the limits of his role in relation to the architect's, even or perhaps especially within an academic context.

In this regard a comparison with Cecil Balmond, a leading international engineer working for Arup, may be useful. Through working with some of the world's leading architects, Balmond has increasingly come to see himself as not only an engineer but an architect as well. In projects such as the Serpentine Pavilion, for example, he and his team at the AGU (Advanced Geometry Unit) have played a leading role in conjunction with a number of international architects. But the question arises: would these projects (and others with a greater degree of programmatic and spatial complexity) have turned out differently without the involvement of the architects? I believe that the answer to such a question would in all likelihood be yes. It is the collaboration that imbues these projects with their varying degrees of success. And it is the recognition of this simultaneous independence and interdependence that makes the creative process between architects and engineers so valuable today. Both Balmond with Arup and Kara with AKT are working in ways that are shaping a new – emergent and experimental – paradigm of contemporary design practice.

It seems that Kara's initial experience at the AA, and more recently his teaching commitments at Cornell and Harvard, have made him, and in turn the whole AKT organisation, even more sensitive to the value of both speculative research and better means of communicating ideas between engineers and architects. AKT are now in great demand amongst many of Europe's most innovative architectural practices, thanks to the combination of the quality of their services and their enthusiasm for complexity. Yet the roots of this form of thinking have been with them from the very beginning, as can be seen, for example, in the development of one of their early projects, the Peckham Library. Together with Alsop as the architects, AKT devised an unconventional solution to the structure of the forecourt of the building, with 'dancing columns' that were at once a response to the technical problem of supporting the projecting upper portion of the building and a way of animating the public space in front. The success of this project has had a dramatic impact on a relatively underprivileged part of London, and has made the use of the library a popular activity far beyond the expectations of the local borough that was responsible for commissioning it in the first place.

But what is unique about the more recent developments within AKT is the way in which the involvement with some of the most groundbreaking forms of architectural education has been literally enfolded into the practice. One of the early moves was the hiring of architects to help interpret (maybe even decipher) the work of architects, not so much for AKT as for the architects themselves. The type of work that is carried out by an architect working within the context of an engineering firm is by definition neither pure engineering nor pure architecture, but rather a form of hybrid practice that through its oscillations between the disciplines produces new forms of knowledge of potential benefit to both. This, for example, is the case with the firm's research on the bridge designed by Future Systems for Land Securities. A free-form envelope twisted around a pair of paths, the bridge connects two floors of separate office buildings with a hexagonal mesh as its cladding system. AKT's research into the process of form-finding, including the development of specific computer programs, helped to integrate the structural and constructional logic and efficiency of the bridge with its aesthetic criteria. This growing emphasis on topological and parametric surfaces that are at once both structure and skin has also opened up new possibilities in the field of construction. A case in point is the use of ship-building technology for the cafe by the beach in Littlehampton that AKT recently completed together with Thomas Heatherwick. This combination of new and old technology, of parametric design and shipbuilding, is itself a recognition of the value of rediscovering traditional technologies in new and unexpected contexts.

In addition to the abstract, rational and experiential knowledge gained through construction, the research work carried out within AKT by both architects and engineers builds on the tacit knowledge acquired by the firm to help its project of pragmatism, a project situated between creativity and problem-solving.

General information of the Phæno Science Centre
see chapter COMPLEXITY, case CONES, page 82

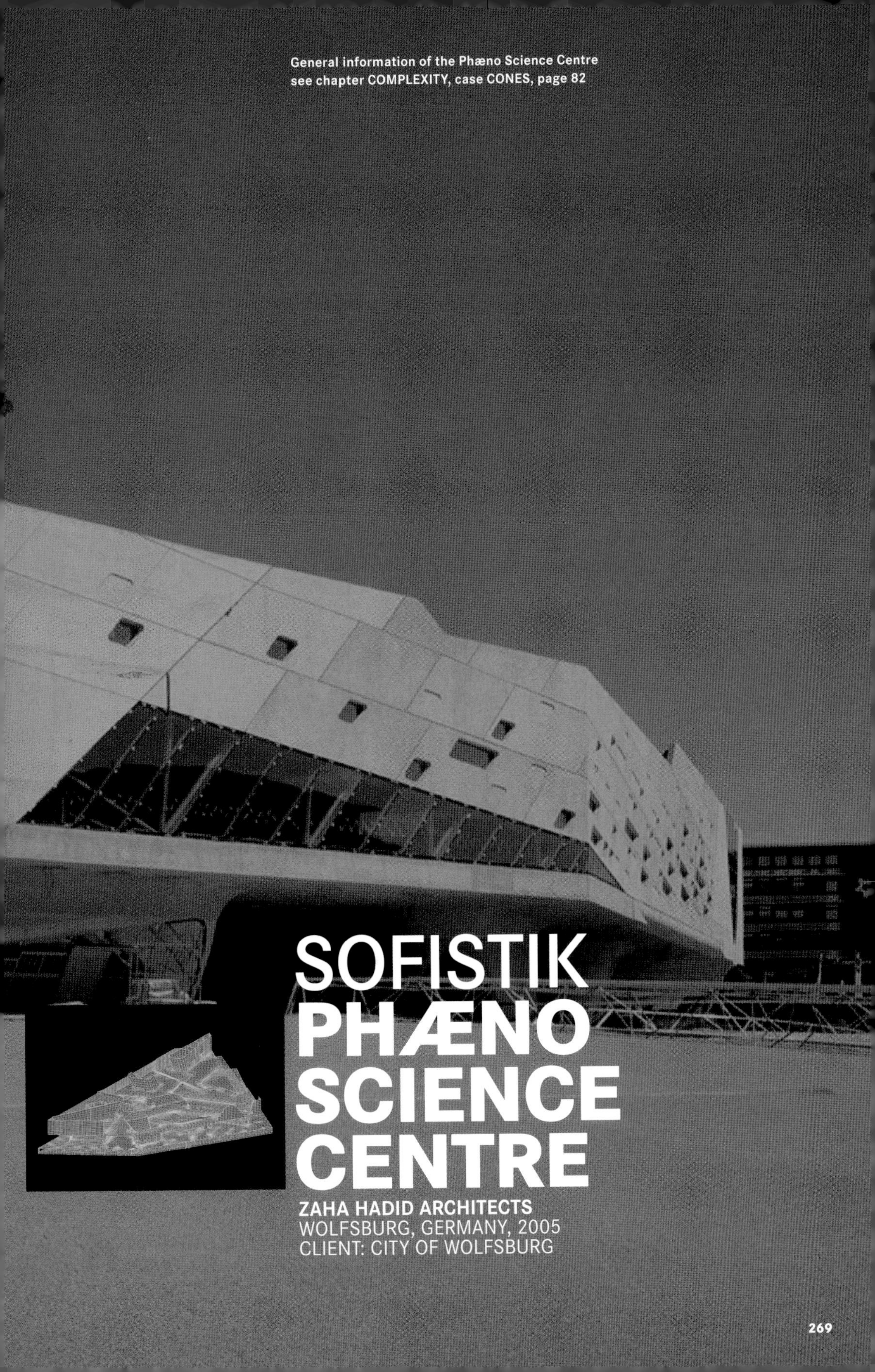

SOFISTIK
PHÆNO SCIENCE CENTRE

ZAHA HADID ARCHITECTS
WOLFSBURG, GERMANY, 2005
CLIENT: CITY OF WOLFSBURG

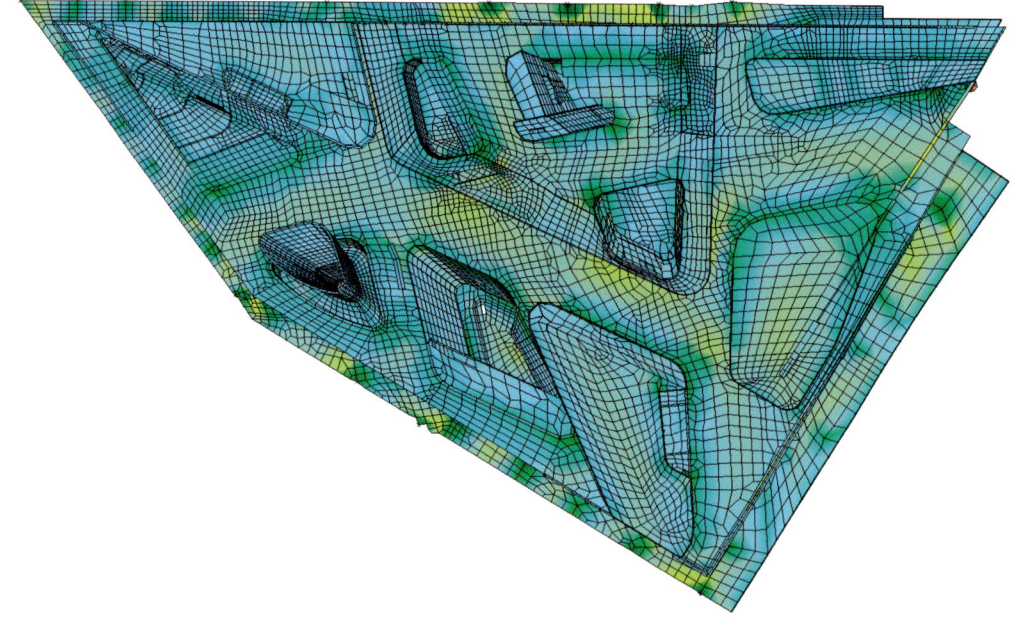

What is the difference in the analysis of the Phæno Science Centre and Strand link bridge?

In the first instance it is about a scale difference. The Phæno Centre is a much larger structure than the bridge, which is a single element. The first step in the Phæno analysis was to get an accurate and stable 3D CAD model; this is then used to build several FEA (Finite Element Analysis) models. For instance, a global model of the 10 cones was built to analyse the behaviour of the ten cones in interaction with the main waffle slab. It was essential to do this as the structure of the waffle slab is a 134m monolith without any movement joints. What we had to simulate is how this behaves when it is cast with the corners each of a varying distance. In addition, we had to check the slab for bending between cones, the most important element being the thermal expansion and contraction of the slab, as this pulls and pushes the cones. Other local models were also built to cross-check the waffle slab, façades and the bridge connecting to the Autostadt.

The global model also then extends to cover the basement and main waffle slab, including the mezzanine levels and the cones, so it contained over 17,000 finite elements. The mesh for these elements was manually created using AutoCAD surface meshes and converting them into finite elements since no mesh generator at the time was able to generate a valid model for such a complex three-dimensional structure. Special adjustment was made to model the material behaviour of the self-compacting concrete accurately since it differs in properties to conventional concrete was used in other parts.

The steel roof is more of a 'line element' and was modelled using a frame analysis package. The steel roof consists of a spatial Vierendeel truss spanning two ways and each note varying, supported only as four cones to create column-free space; again a generator was not used.

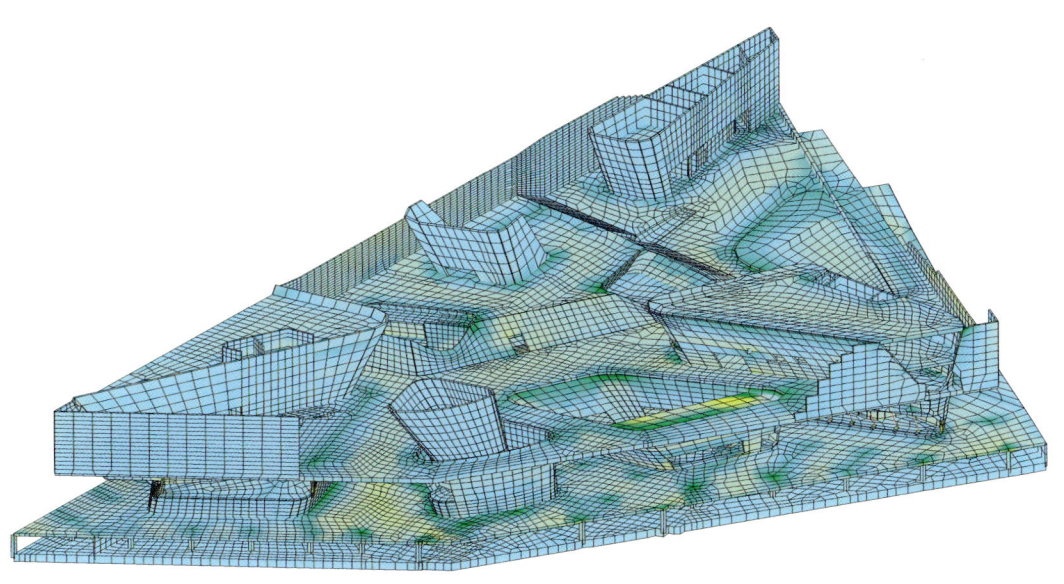

The Land Securities bridge is much more advanced in the way it is analysed and optimised. The Sofistik analysis is used in a homogenous geometrical process where computation techniques connect the result back to the geometrical model as a feedback, generating optimised or preferred meshes automatically. In that way the combination of hexagonal and triangular tiles comes from this optimised feedback. The new tools provide feedback at a great speed.

Given what you discussed earlier, how do you divide the global models into local models? Do you rely on the analysis provided by one tool only?

The division of the model requires 'human transaction' by understanding the bottom-up sequence and possible 'pieces' the construction would take place in. This is easiest to relate to the cones which are at their most unstable state when they have been constructed up to the underside of the main floor. Here they behave individually, so it is a moment in the analysis that has to be modelled and tested. Similarly, as the horizontal slab is constructed in stages and each step is analysed, it's only when the slab is cast fully and curved that the cones act as a group and are stable. In terms of using other software, there simply isn't the time to do that... But the engineers undertake hand calculations to approximate the behavior of the analysis prior to progressing to the next step.

How do you collaborate with Sofistik to advance the software?

It is a very interesting point. All good software writers are always seeking to test and push their software. To get the best results, the expertise of both partners in this collaboration is essential; a good engineering understanding and a similarly good understanding of the software are essential. The challenge of the model motivated the software writer as much as it did us, as most advances in technology need high levels of motivation and skill. The package was continually refined throughout the project to improve speed, visualisation and to cope with complex load combinations.

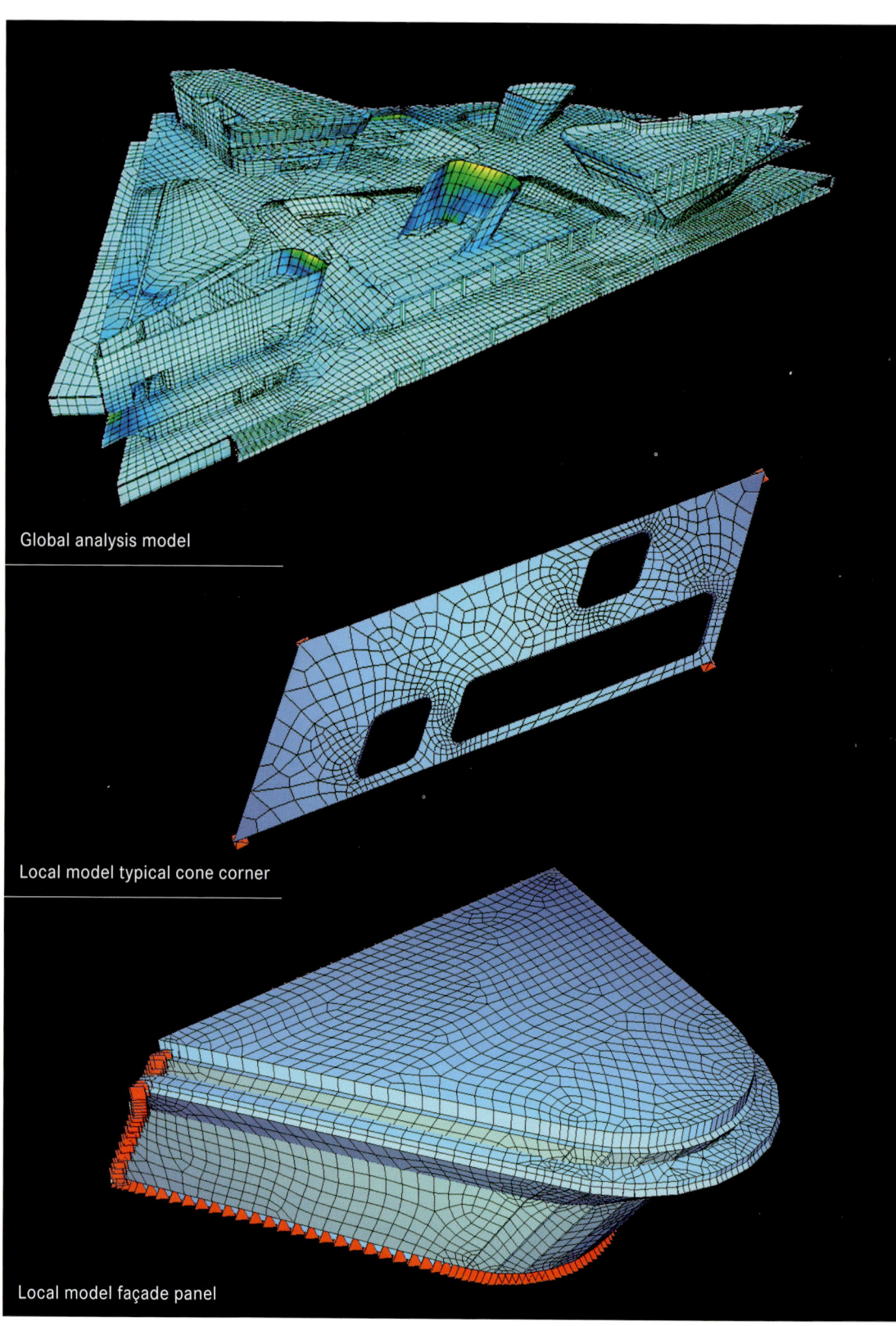

Global analysis model

Local model typical cone corner

Local model façade panel

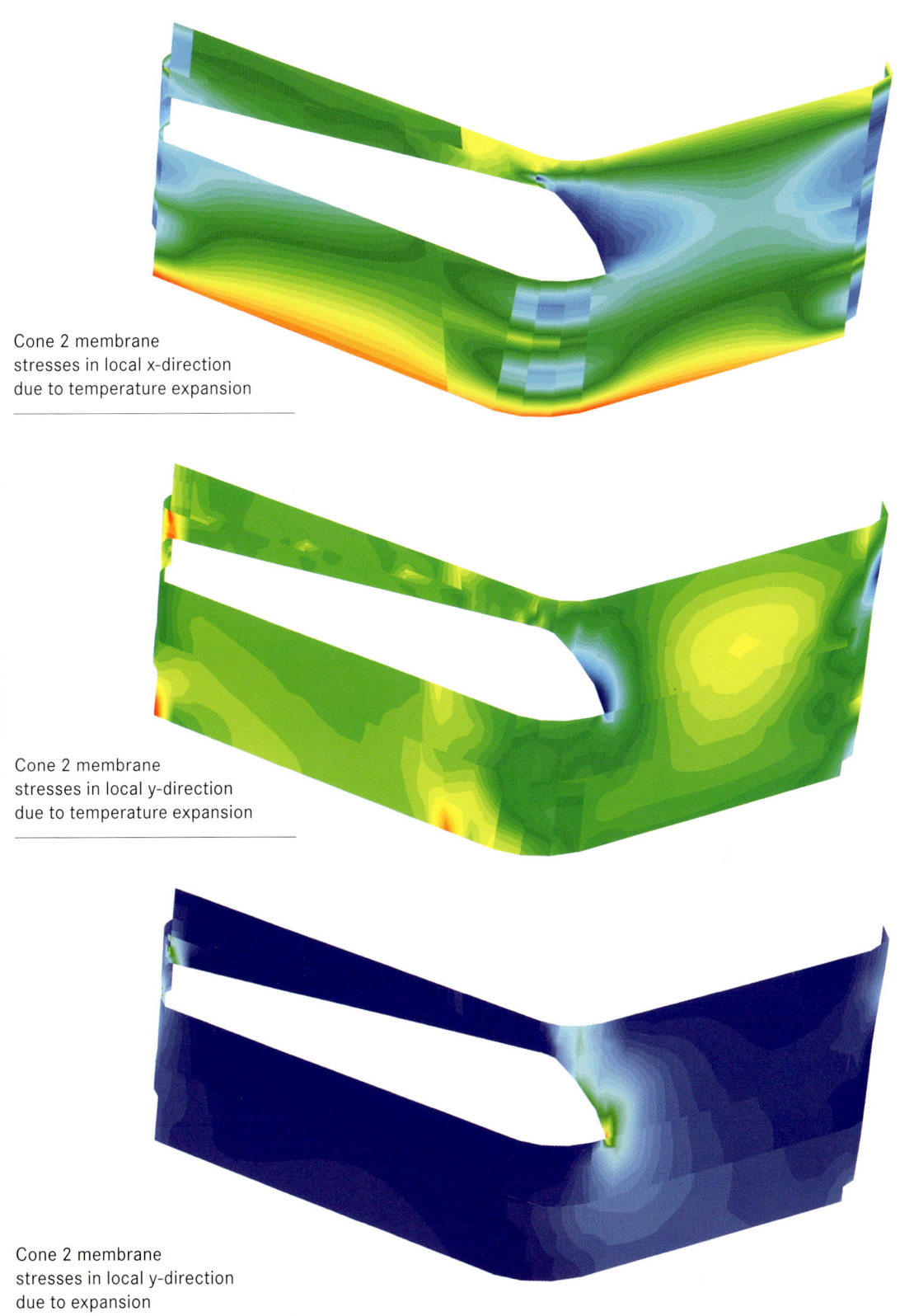

Cone 2 membrane
stresses in local x-direction
due to temperature expansion

Cone 2 membrane
stresses in local y-direction
due to temperature expansion

Cone 2 membrane
stresses in local y-direction
due to expansion

AKT collaborated with George Legendre of IJP Architects, a teaching colleague from the Architectural Association, on a winning competition entry for a 250 metre-long pedestrian and cycle bridge through parkland in Singapore. The Southern Ridges Bridge design proposes an expressive steel superstructure spanning concrete pylons, whose height varies in response to the changing ground level, topped by a timber deck. Integrating engineering and architecture, this approach ensured complete consistency between the original idea and construction method, because all the information necessary for design, component manufacture, fabrication and erection is deduced from the same formula.

ALGORITHM, FORM GENERATION SOUTHERN RIDGES BRIDGE

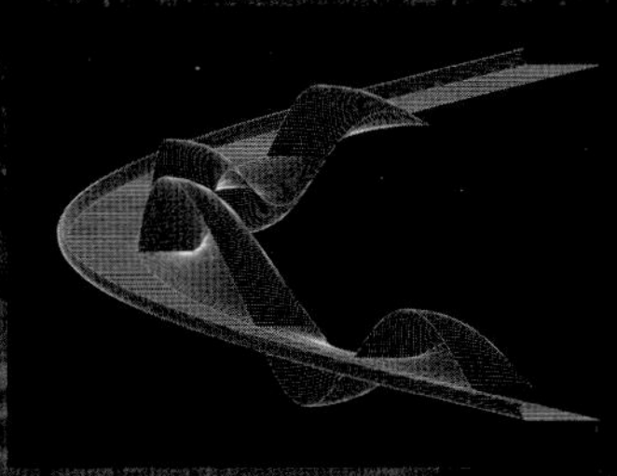

IJP ARCHITECTS / RSP ARCHITECTS
SOUTHERN RIDGES, SINGAPORE, 2004–2008
CLIENT: URBAN REDEVELOPMENT AGENCY

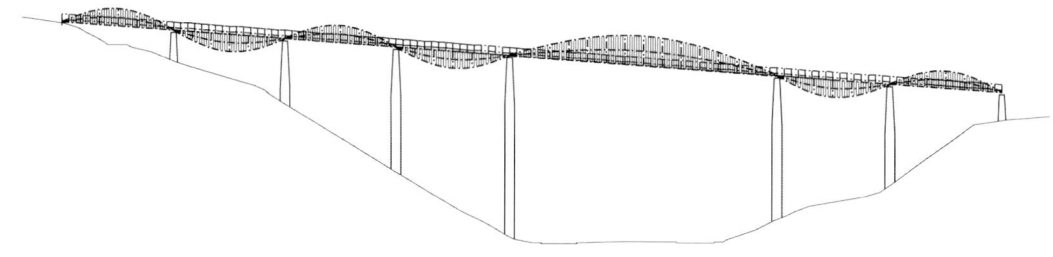

$$\text{density\_of\_threads}:=10 \quad \text{length\_of\_bridge}:=270 \quad \text{length\_of\_segment}:=72 \qquad \text{beginning of interval}:= \frac{0 \ \pi}{100}$$

$$\text{last\_in}:=\text{density\_of\_threads}\cdot\frac{\text{length\_of\_segment}}{0.5\cdot10} \quad \text{last\_jn}:=5\cdot\text{density\_of\_threads}$$

$$\text{number\_of\_pockets}:=3 \qquad\qquad\qquad \text{deltaH}:=20.5 \qquad\qquad\qquad \text{periodic\_noise\_coefficientTRANS}:=1$$

$$\text{width\_of\_segments}:=3.5 \qquad\qquad\qquad \text{jn}:=0,1..\text{last\_jn} \qquad\qquad\qquad \text{periodic\_noise\_coefficientLONG}:=1$$

$$\text{inflection\_of\_segments}:=26 \qquad\qquad \text{in}:=0,1..\text{last\_in} \qquad\qquad\qquad \text{mirror\_surface}:=-1$$

$$\qquad\qquad\qquad\qquad\qquad\qquad\qquad\qquad\qquad\qquad\qquad\qquad\qquad \text{mirror\_pockets}:=-1$$

$$\text{height\_of\_segment}:=3.5\cdot\text{concavity\_sign} \qquad \text{fraction\_of\_trigo\_arc}:= \frac{\text{length\_of\_segment}\cdot100}{\text{length\_of\_bridge}} \qquad \text{concavity\_sign}:=1$$

$$\qquad\qquad\qquad\qquad\qquad\qquad\qquad\qquad\qquad\qquad\qquad\qquad\qquad\qquad \text{depth\_of\_pocket}:=1$$

$$\text{height\_of\_segment}:=3.5\cdot\text{concavity\_sign}$$

$$\text{Bend}_{in,jn} :=\sin\left(\text{beginning\_of\_interval}+\frac{in}{\text{last\_in}}\cdot\pi\cdot\frac{\text{fraction\_of\_trigo\_arc}}{100}\right)\cdot\text{inflection\_of\_segment}$$

$$\text{slopeA}_{in,jn} :=in\cdot\frac{\text{deltaH}\cdot\text{fraction\_of\_trigio\_arc}}{100\cdot\text{last\_in}}$$

$$\text{jThread}_{in,jn} :=\text{mirror\_surface}\cdot\text{mirror\_pockets}\left[\sin\left[\frac{jn}{\text{last\_jn}}\cdot\left(\pi\,\text{depth\_of\_pocket}\right)\right]\cdot\text{width\_of\_segment}+\text{mirror\_pockets}\cdot\text{Bend}_{in,jn}\right]$$

$$\text{iThread}_{in,jn} :=1\cdot\left(\frac{in}{\text{last\_in}}\cdot\text{length\_of\_segment}\right)$$

$$\text{shape}_{in,jn} :=\left[\sin\left(\frac{in}{\text{last\_in}}\cdot\text{number\_of\_pockets}\cdot\pi\right)\cdot\left(\cos\left(\frac{jn}{\text{last\_jn}}\cdot\pi\right)+-1\right)\right]\cdot\frac{\text{height\_of\_segment}}{2}+\text{slopeA}_{in,jn}$$

A formula to make a bridge

The process employed on this project was another example of how approach is just as important as the tools you use.

Although the project made the best use of the software tools to date, it was initially conceived with a sheet of paper and some creative thought which set down the basic design parameters for the structure and architecture of a multi span bridge.

With this information a MathCAD formula was developed which defined both the skeleton and surface of the bridge. Unlike many projects this gave both IJP and AKT the benefit of having complete control over the geometry of the project from the early stages using only one formula. On a practical level this meant that the geometry could be moved between software platforms with relative ease and could be exported in a resolution to suit the different tasks required. On a strategic level the control of such a form meant that the design was being progressed with confidence on the outcome of all conditions.

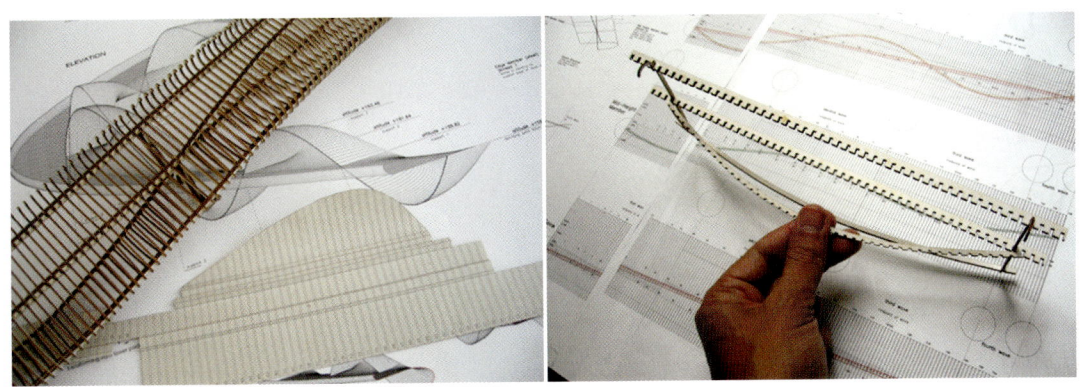

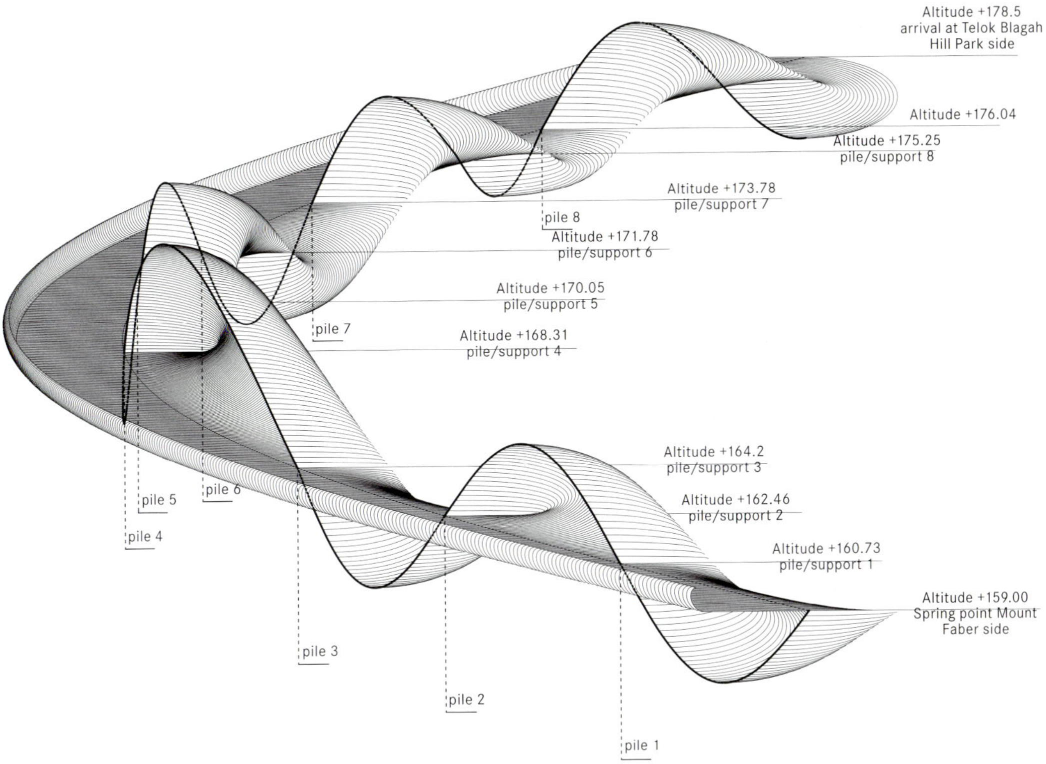

Altitude +178.5
arrival at Telok Blagah
Hill Park side

Altitude +176.04

Altitude +175.25
pile/support 8

Altitude +173.78
pile/support 7

pile 8
Altitude +171.78
pile/support 6

Altitude +170.05
pile/support 5

pile 7
Altitude +168.31
pile/support 4

Altitude +164.2
pile/support 3

pile 5   pile 6
Altitude +162.46
pile/support 2

pile 4
Altitude +160.73
pile/support 1

Altitude +159.00
Spring point Mount
Faber side

pile 3

pile 2

pile 1

Elevation of the bridge

The IJP model was imported into our frame analysis software with design loads and member properties added which could inform the early stages of design. This enabled a real-time response to the development of the MathCAD formula and the design of the bridge. The intelligent feedback in the early stages of design may not be unusual but the resolution of detail certainly was.

In addition long span structures such as bridges bring with them different problems not just with static loading (such as dead loads) but also live loading and their dynamic response. The sensitivity of such structures to the dynamic response is usually a combination of many factors some of which don't get solved until the later stages of design, but in this instance we were able to use this data to refine the original seed.

Mesh density and form variables TEMPLATE 3

density := 4 last_in := 10·density last_jn := 10·density   lengthFactor := 1   mesh density factor

in := 0, 1 .. last_in   jn := 0, 1 .. last_jn   widthFactor := 0.5   bounds of mesh range

periods_longitudinal := 1periods_transversal := 1.9   heightFactor := 0.4   mesh range

number of periods along in and jn

range of background mesh

Parametric definitions of the meshes

$\text{iThread}_{in, jn} := \dfrac{in}{last\_in} \cdot lengthFactor$ longitudinal distribution   **non uniform** spacing of iThreads

$\text{jThread}_{in, jn} := \sin\left(\dfrac{jn}{last\_jn} \cdot periods\_transversal \cdot \pi\right) \cdot widthFactor$ **free range**

$\text{shape}_{in, jn} := \left[\sin\left(\dfrac{in}{last\_in} \cdot periods\_longitudinal \cdot \pi\right) \cdot \left(\cos\left(\dfrac{jn}{last\_jn} \cdot periods\_transversal \cdot \pi\right)\right)\right] \cdot heightFactor$

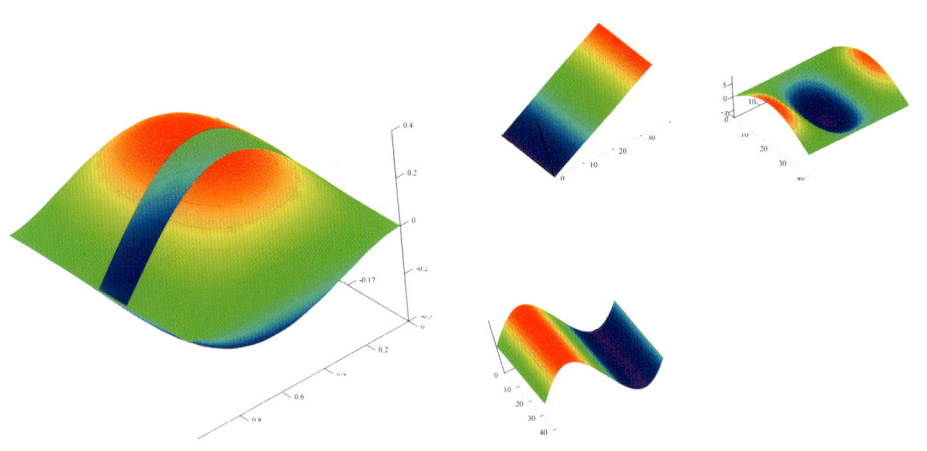

Mesh density and form variables TEMPLATE 3

density := 4 last_in := 10·density last_jn := 10·density   lengthFactor := 2   mesh density factor

in := 0, 1 .. last_in   jn := 0, 1 .. last_jn   widthFactor := 0.5   bounds of mesh range

periods_longitudinal := 2periods_transversal := 1.9   heightFactor := 0.4   mesh range

number of periods along in and jn

range of background mesh

Parametric definitions of the meshes   **Copyright IJP Corporation 2004-06**

$\text{iThread}_{in, jn} := \dfrac{in}{last\_in} \cdot lengthFactor$ longitudinal distribution   **non uniform** spacing of iThreads

$\text{jThread}_{in, jn} := \sin\left(\dfrac{jn}{last\_jn} \cdot periods\_transversal \cdot \pi\right) \cdot widthFactor$ **free range**

$\text{shape}_{in, jn} := \left[\sin\left(\dfrac{in}{last\_in} \cdot periods\_longitudinal \cdot \pi\right) \cdot \left(\cos\left(\dfrac{jn}{last\_jn} \cdot periods\_transversal \cdot \pi\right)\right)\right] \cdot heightFactor$

Parametric Plot

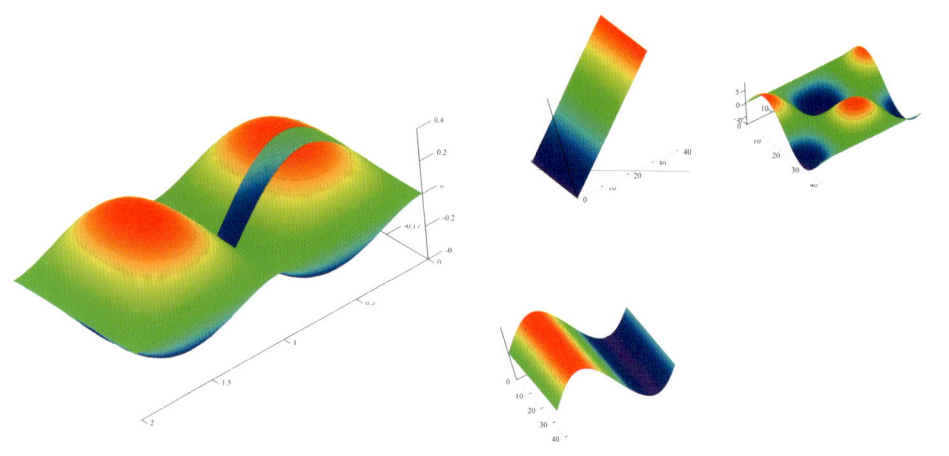

All images provided by ITP

density := 4 last_in := 10·density last_jn := 10·density   lengthFactor := 3   mesh density factor

in := 0, 1 .. last_in   jn := 0, 1 .. last_jn   widthFactor := 0.5   bounds of mesh range

periods_longitudinal := 3periods_transversal := 1.9   heightFactor := 0.4 mesh range

number of periods along in and jn

range of background mesh

Parametric definitions of the meshes   **Copyright IJP Corporation 2004-06**

$$iThread_{in,jn} := \frac{in}{last\_in} \cdot lengthFactor$$ longitudinal distribution **non uniform** spacing of iThreads

free range

$$jThread_{in,jn} := \sin\frac{jn}{last\_jn} \cdot periods\_transversal \cdot \pi \cdot widthFactor$$

$$shape_{in,jn} := \left[\sin\frac{in}{last\_in} \cdot periods\_longitudinal \cdot \pi\right) \cdot \left(\cos\frac{jn}{last\_jn} \cdot periods\_transversal \cdot \pi\right] \cdot heightFactor$$

Parametric Plot

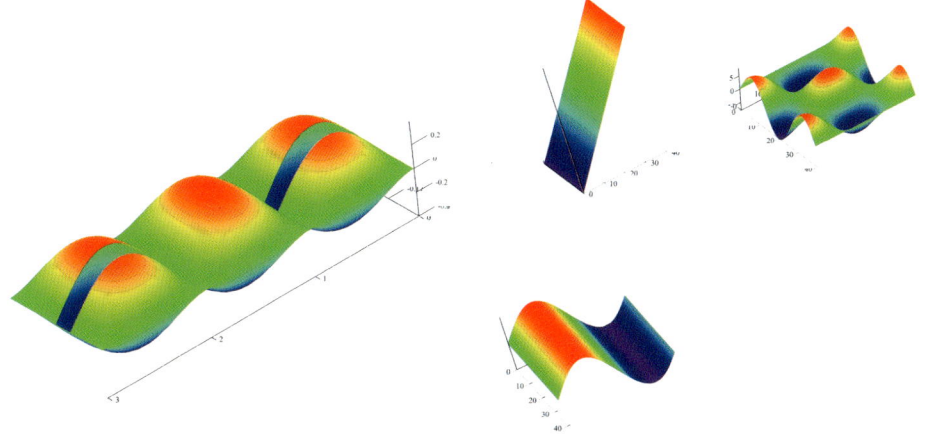

---

density := 4 last_in := 10·density last_jn := 10·density   lengthFactor := 3   mesh density factor

in := 0, 1 .. last_in   jn := 0, 1 .. last_jn   widthFactor := 0.5   bounds of mesh range

periods_longitudinal := 3periods_transversal := 1   heightFactor := 0.4 mesh range

number of periods along in and jn

range of background mesh

Parametric definitions of the meshes   **Copyright IJP Corporation 2004-06**

$$iThread_{in,jn} := \frac{in}{last\_in} \cdot lengthFactor$$ longitudinal distribution **non uniform** spacing of iThreads

free range

$$jThread_{in,jn} := \sin\frac{jn}{last\_jn} \cdot periods\_transversal \cdot \pi \cdot widthFactor$$

$$shape_{in,jn} := \left[\sin\frac{in}{last\_in} \cdot periods\_longitudinal \cdot \pi\right) \cdot \left(\cos\frac{jn}{last\_jn} \cdot periods\_transversal \cdot \pi\right] \cdot heightFactor$$

Parametric Plot

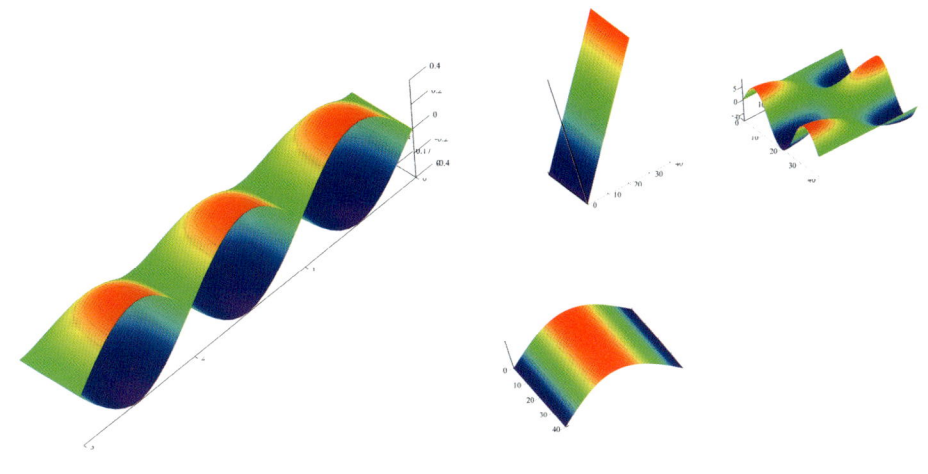

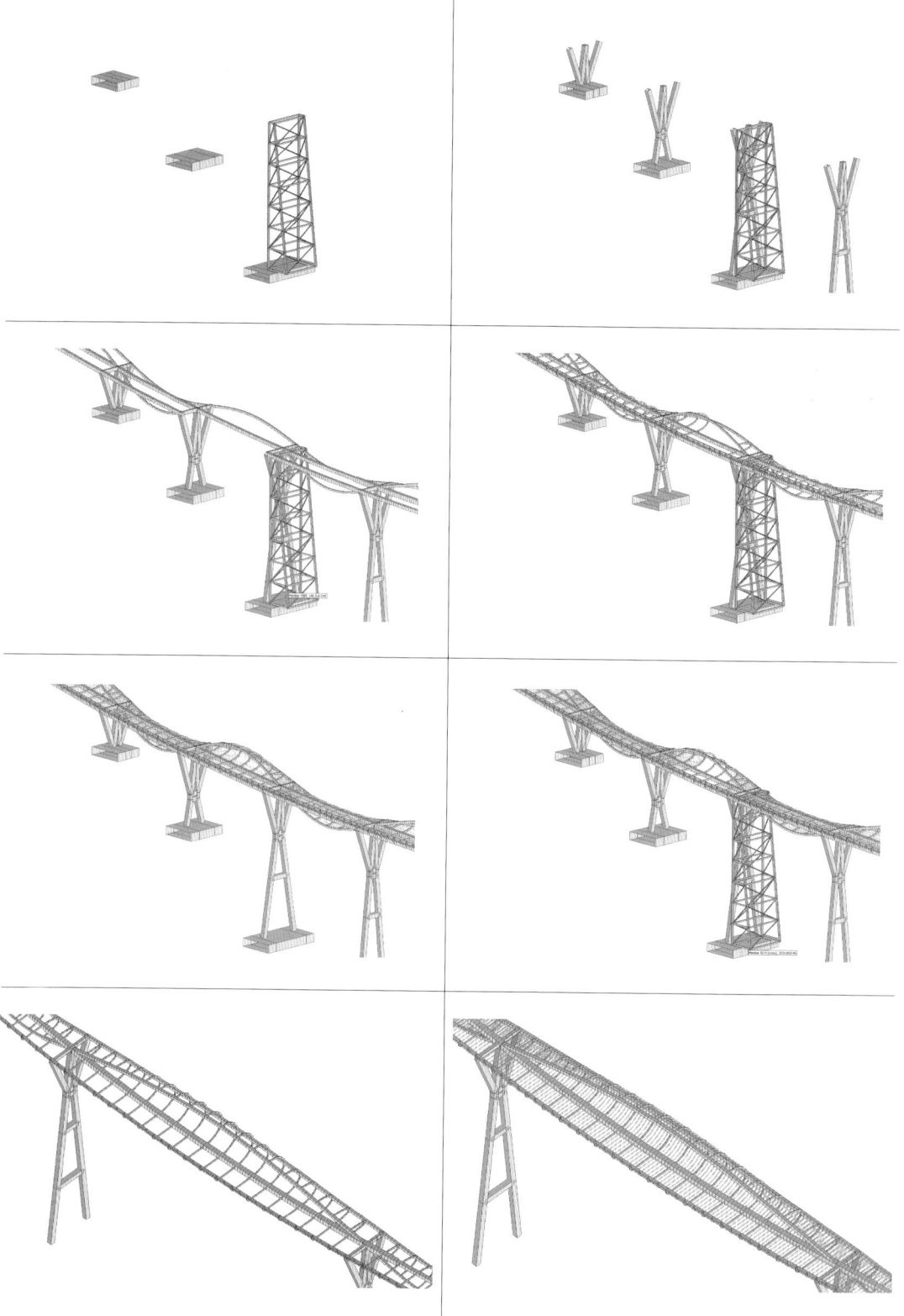

31/03/2007

02/04/2007

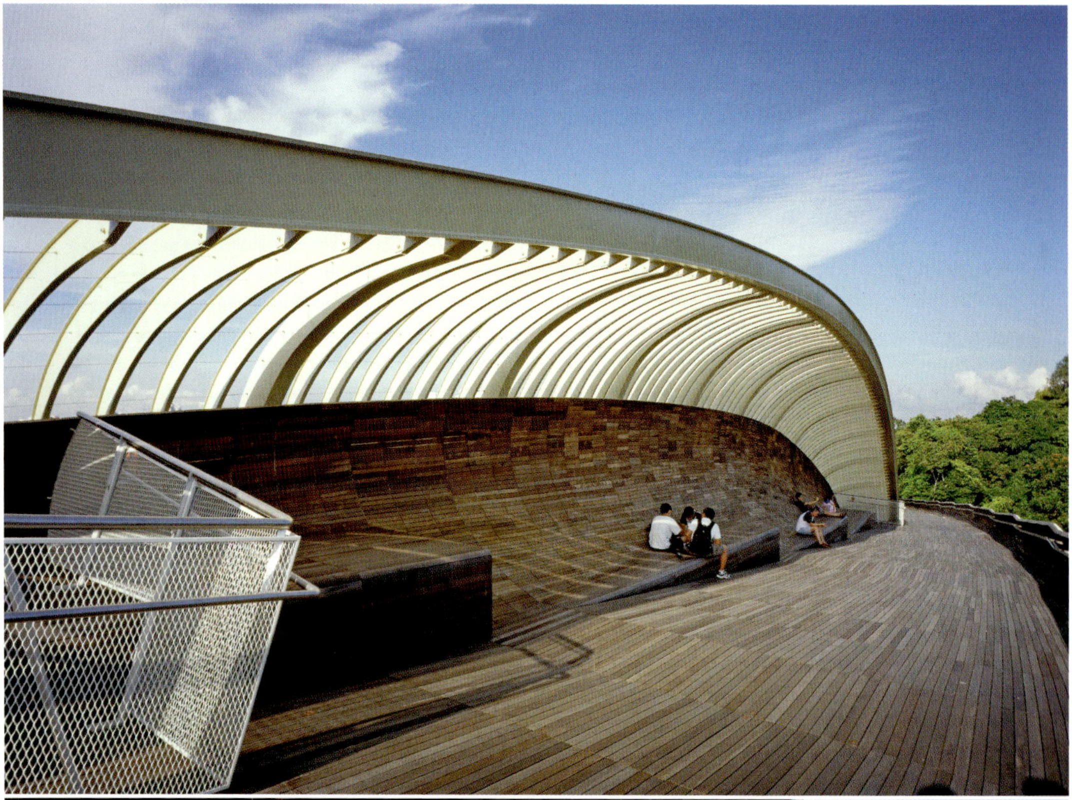

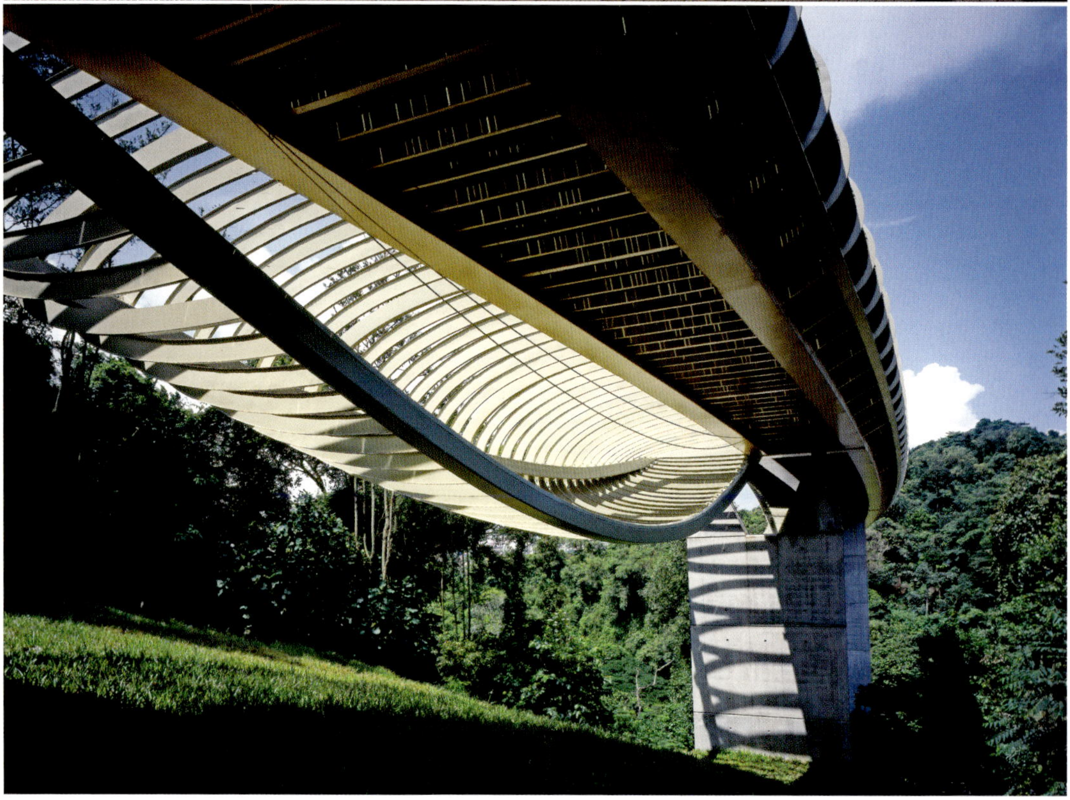

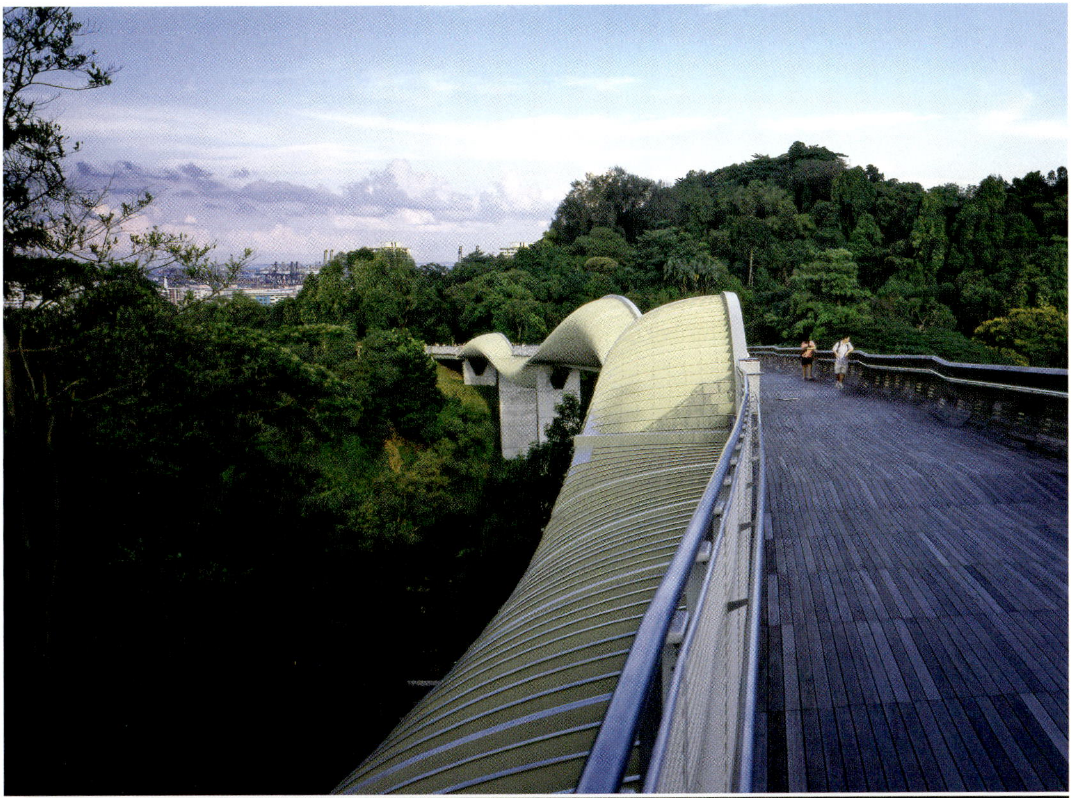

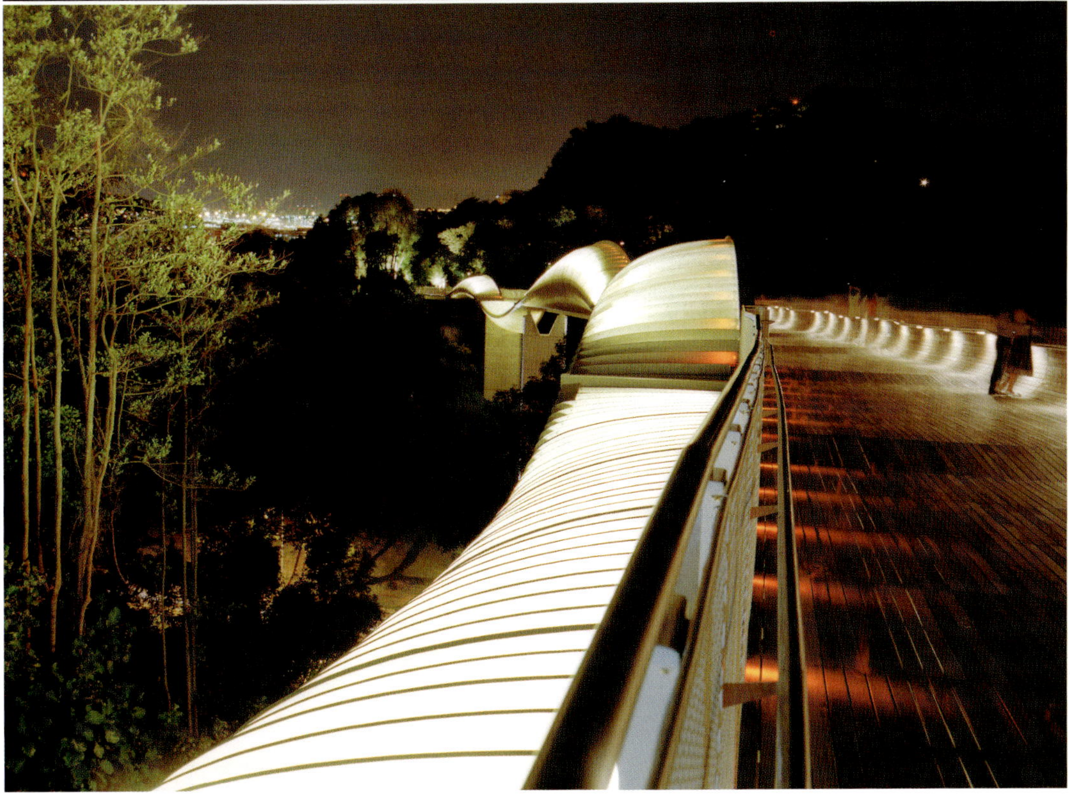

As the design developed so were we able to consider particular types of member profiles for the bridge and give consideration to their manufacture and fabrication, therefore engaging the wider section of the industry that normally comes onboard at a later stage in the project.

For us this process illustrated well the responsibility that comes with developing forms and surfaces. The design team has to be in control of the process, otherwise there is a danger that the design is never de-risked in terms of technical input, manufacture, fabrication, delivery and programme. In other words, once you embark on this path, at the very least, it needs a commitment to follow the process through, but more importantly it also requires a clear strategy, whether using software tools or not.

Tools used without thought will only create chaos!

The case study of the Land Securities Bridge was developed by focusing on the feasibility study undertaken by Adams Kara Taylor in supporting the design concept by Future Systems for Land Securities.

The Land Securities Bridge is designed as a free-form envelope that twists around a pair of paths, connecting two floors of two office buildings with a hexagonal mesh that forms its cladding system. The complex structure is realised as a lattice of nodes and struts supporting lightweight polycarbonate panels. Guidance from the fabrication industry suggested that the number of nodes in the frame would be the biggest cost factor, as the connections are to be fabricated one by one, welding a machined spherical piece to standard cut tubes.

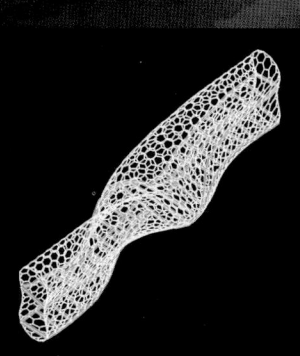

HEXAGONAL PANELIZATION
STRAND LINK BRIDGE

FUTURE SYSTEMS
LONDON, 2005
CLIENT: LAND SECURITIES

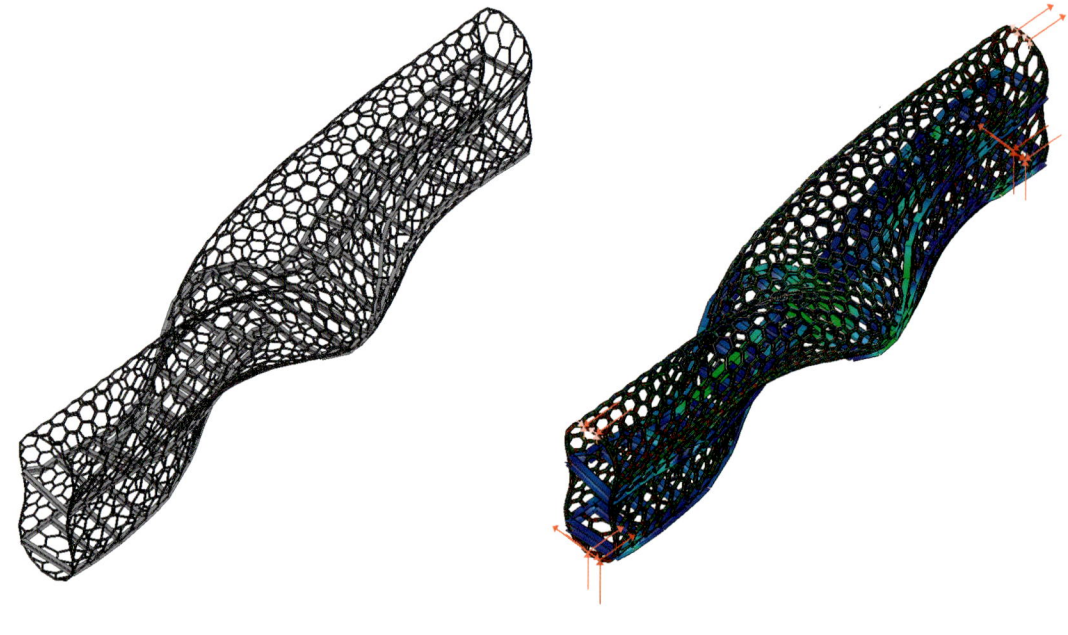

At this stage in our work, we were beginning to get more and more interested in complex geometry. The architect, Future Systems, wanted to respond to the brief — connecting two buildings — via a very free form surface. To some, it might be considered shape-making for the sake of shape, but there is a very interesting logic in terms of going from a horizontal to a vertical shape inside. Future Systems is very clear about their desired end results, so it's great to work with them. They are also very open.

We proposed to look at the concepts of complexity and simplexity; often things that appear complex initially, seem simple later on. We took the shape as it was given and resisted the temptation to straighten it up, in order to see what we could find within that shape that would suggest a structural trajectory.

We began by taking the form and tiling it with a series of known components of hexagons and triangles. We imagined that we would either compose a repetitive module — not a repetitive tile, but a repetitive group of things — or make one universal connection that then translates the geometry in a complex way (a bit like Jubilee Line extension for example >see page 90). Working with computational designers we could test things and calibrate the benefit of one against the other. So we were able to take both triangle and hexagon, and then calibrate different sizes quickly to get an idea of the range that would work on a very complex shape. Often these processes aren't completed due to time constraints, but in this case, through the calibration of sizes and tile types, we were able to find an optimum size and combination to create a module. Rather than just rely on shape or material, we were able to connect it to analytical software. So the real discussion on how to optimise the material is informed by structural analysis, not purely by tiling or fabrication materials.

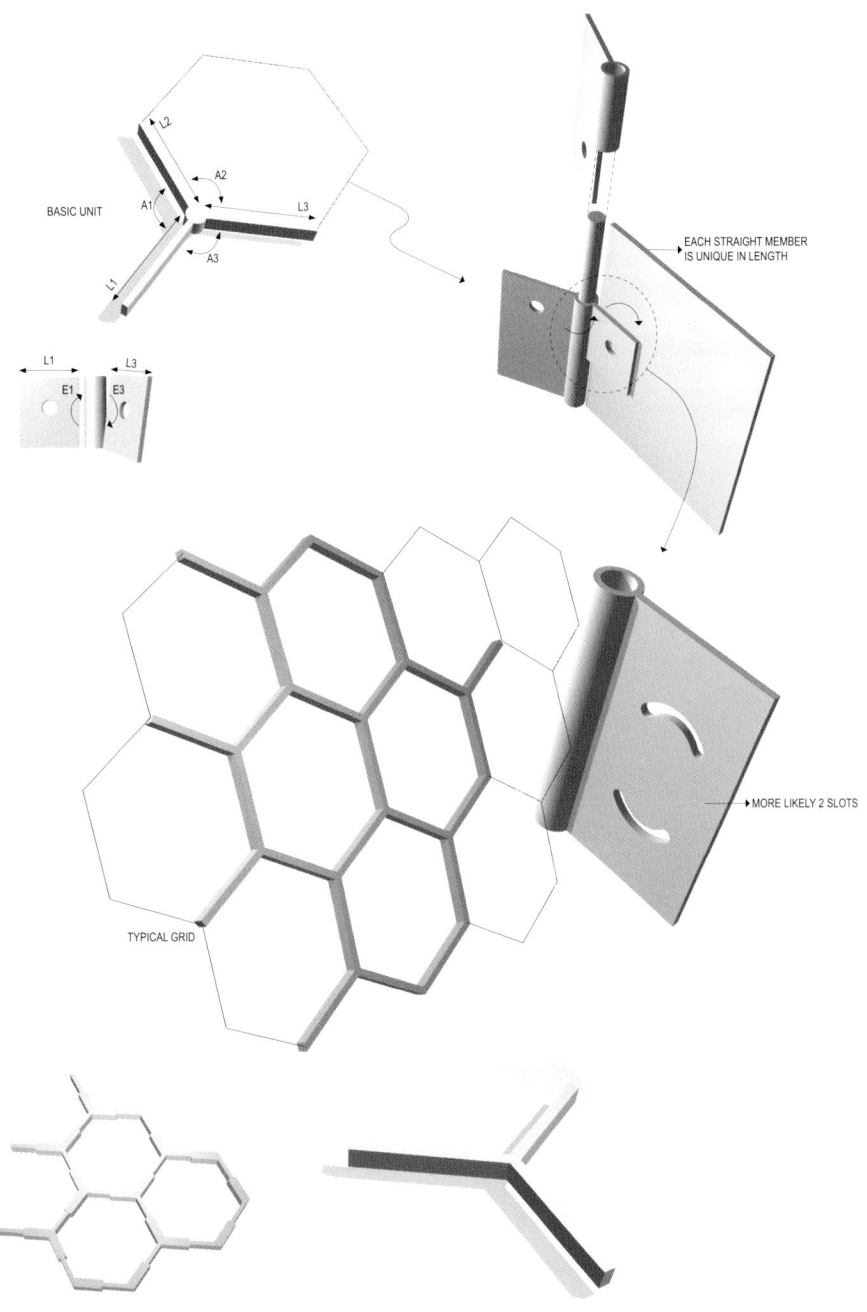

BASIC UNIT

L2

A2

A1

L3

A3

L1

L1

L3

E1

E3

EACH STRAIGHT MEMBER
IS UNIQUE IN LENGTH

MORE LIKELY 2 SLOTS

TYPICAL GRID

Hinge Concept

Effectively, this calibration process allows us to weigh the bridge, to assess lightness and heaviness, and it is also a way of checking costs. If the project had been made of steel, it would have been slightly heavier with more components and would have cost more. In this case, we have achieved a second iteration on how to differentiate a structure through using computational techniques. We were also able to rapidly prototype the stress patterns.

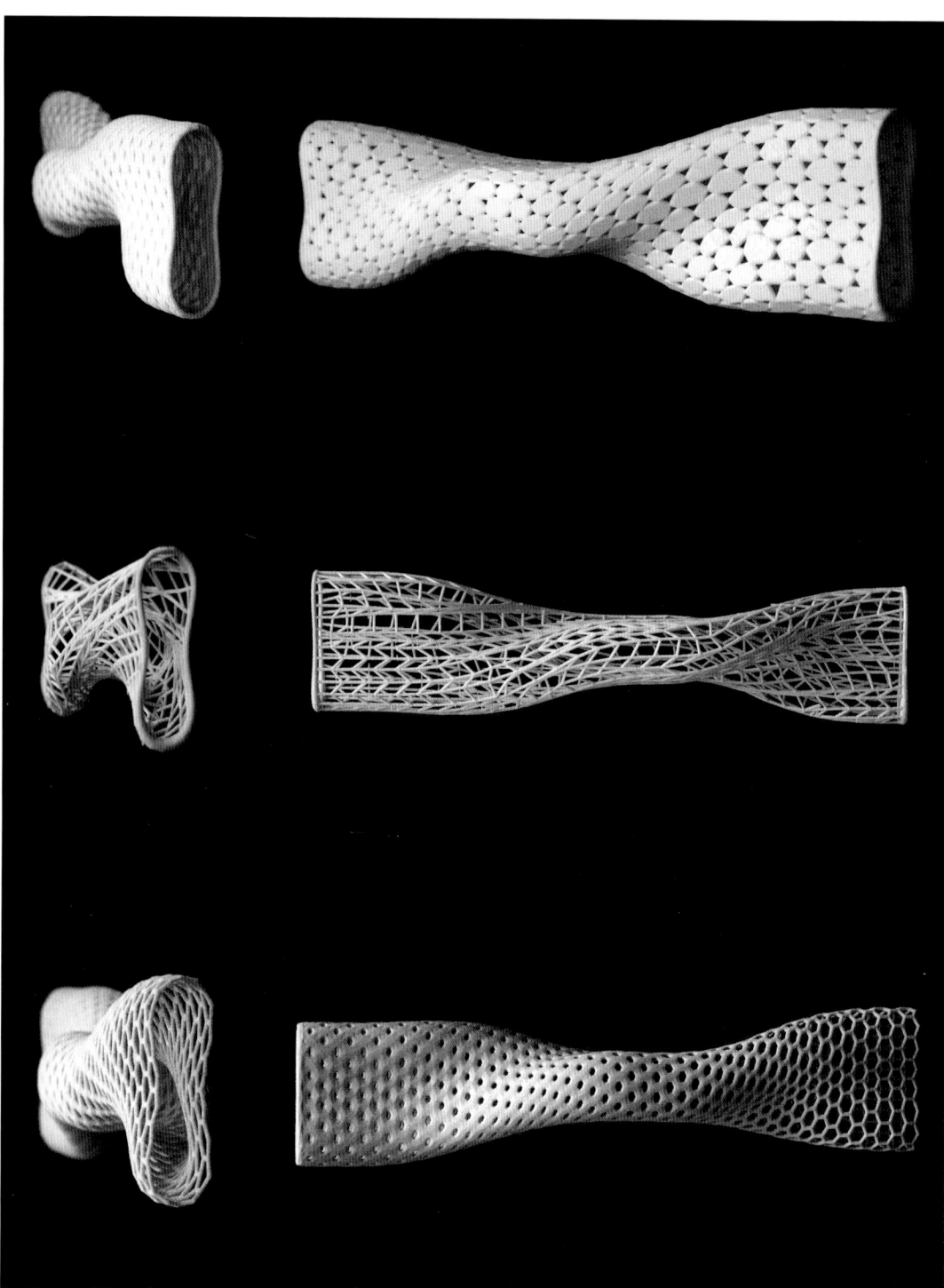

We believe we came to what was probably the most optimum solution in terms of both structure and fabrication. However the client, Land Securities, who is the largest developer in the UK, still hasn't built it, the bridge is located at their headquarters in London.

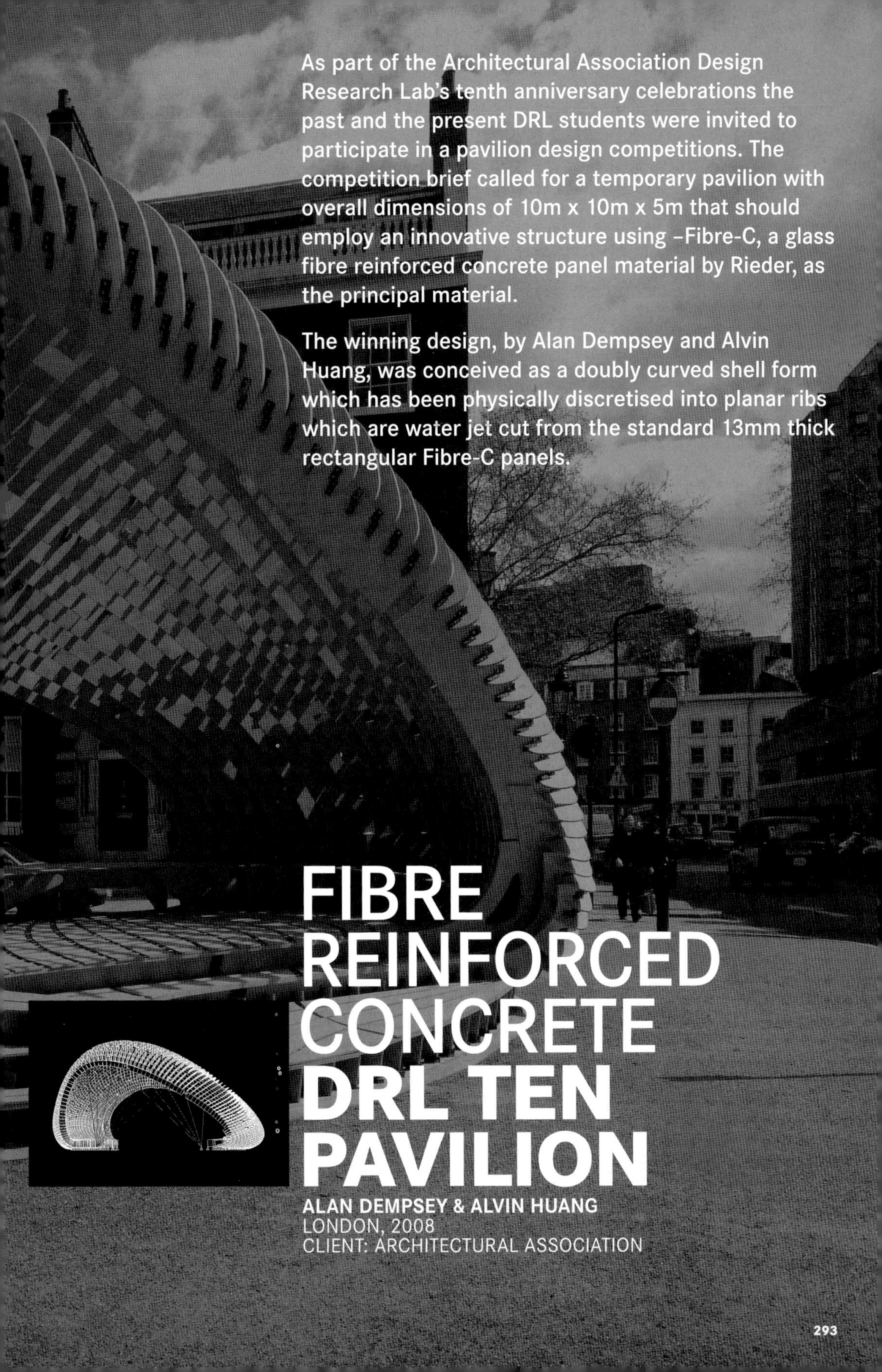

As part of the Architectural Association Design Research Lab's tenth anniversary celebrations the past and the present DRL students were invited to participate in a pavilion design competitions. The competition brief called for a temporary pavilion with overall dimensions of 10m x 10m x 5m that should employ an innovative structure using –Fibre-C, a glass fibre reinforced concrete panel material by Rieder, as the principal material.

The winning design, by Alan Dempsey and Alvin Huang, was conceived as a doubly curved shell form which has been physically discretised into planar ribs which are water jet cut from the standard 13mm thick rectangular Fibre-C panels.

FIBRE REINFORCED CONCRETE
DRL TEN PAVILION

ALAN DEMPSEY & ALVIN HUANG
LONDON, 2008
CLIENT: ARCHITECTURAL ASSOCIATION

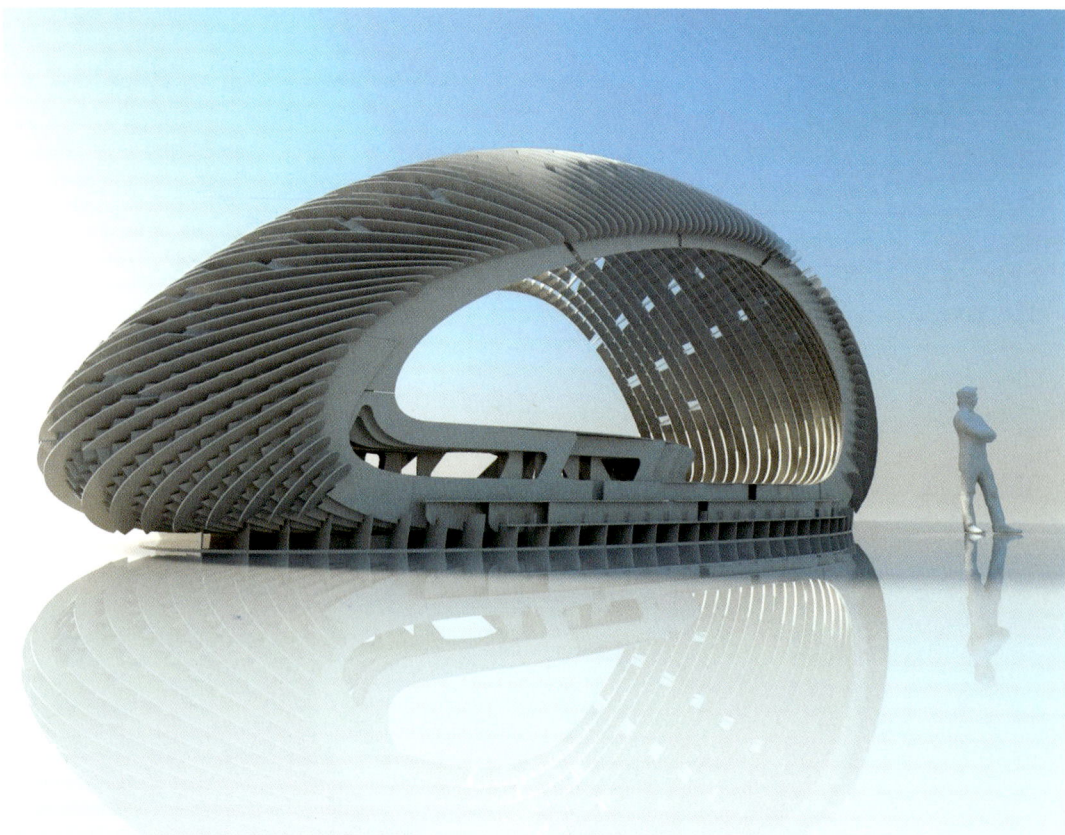

The aspects of this project requiring a particular focus from AKT, derived from both the onerous design programme and the innovative use of material, were:

- The necessity to achieve a particularly fast rate of iteration through the design development.
- The use of physical testing to find the detailed mechanical properties of the two main materials.
- Creation of detailed brick element analysis models in order to mimic the action of the rib to rib connection.

Design Development The discretised and discontinuous shell concept for the pavilion was developed in some detail prior to any involvement from AKT. As a result it was necessary for AKT to find firstly an efficient method of working with the geometrical model from the designers which permitted the structural analysis model to be repeatedly generated quickly as the design developed and secondly to identify at an early stage those aspects of the proposal which should be modified to attain the required structural performance while retaining the overall aesthetic design intent.

For AKT this presented an opportunity to connect digital design with digital fabrication whilst simultaneously testing the students' capacity of learning from 'constructing' the idea. At the same time the 'analysis' and 'testing' at speed illustrated new nonlinear processes in design and construction.

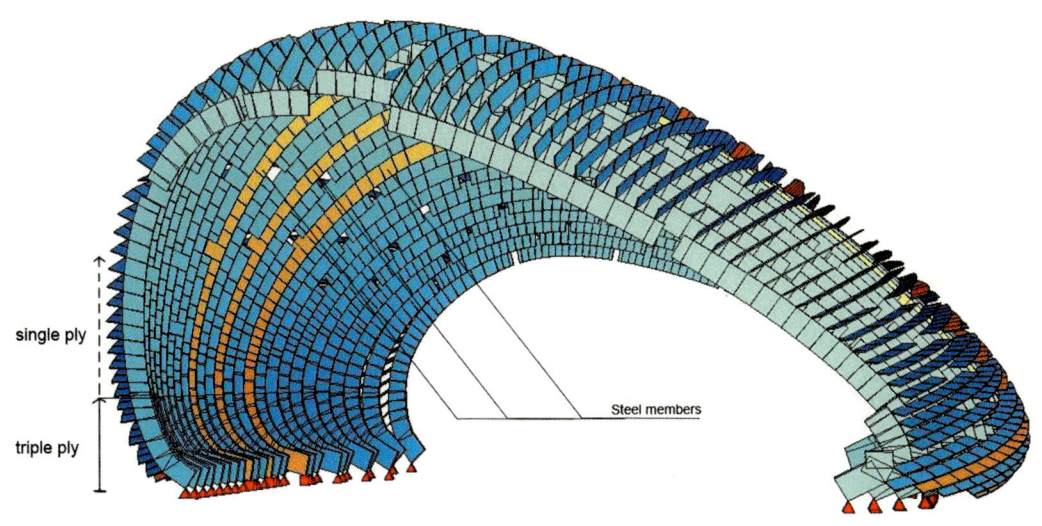

single ply

triple ply

Steel members

Rib to Rib Connection Analysis

In a drive towards achieving a reasonable level of buildability and in recognition of the criticality of construction tolerances in a project of this nature an unusual rib connection detail was proposed in the competition design which employed rubber gaskets at the interface primary and secondary ribs. In terms of the overall structural integrity of the pavilion these connections are required to carry out two main tasks.

Firstly, as described above, the overall front to back stability is derived from the transfer of minor axis bending between the primary and secondary ribs, this minor axis bending is transferred via the rubber gaskets.

Secondly the primary and secondary ribs are discontinuous along their length, this dictates that the load within each rib must be distributed laterally in order that the loads are shared by multiple ribs. In this way, due to the staggered nature of the rib discontinuities, a pseudo continuous overall form is achieved, the gaskets must perform this load transfer function.

Clearly the rotation and shear stiffness of the rib to rib connection is critical to the overall structural performance. This stiffness was assessed using a combination of testing physical models and digital models.

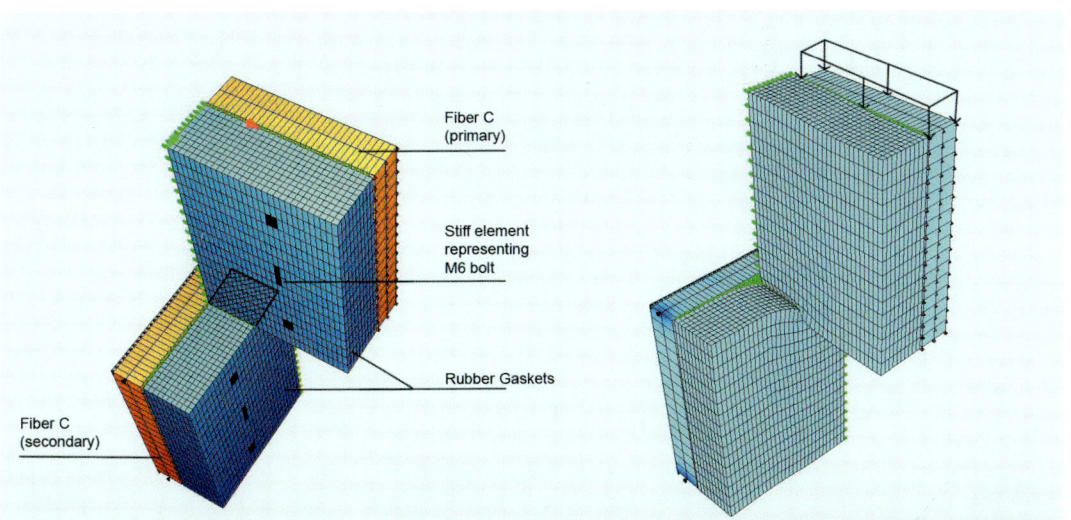

Fiber C
(primary)

Stiff element
representing
M6 bolt

Rubber Gaskets

Fiber C
(secondary)

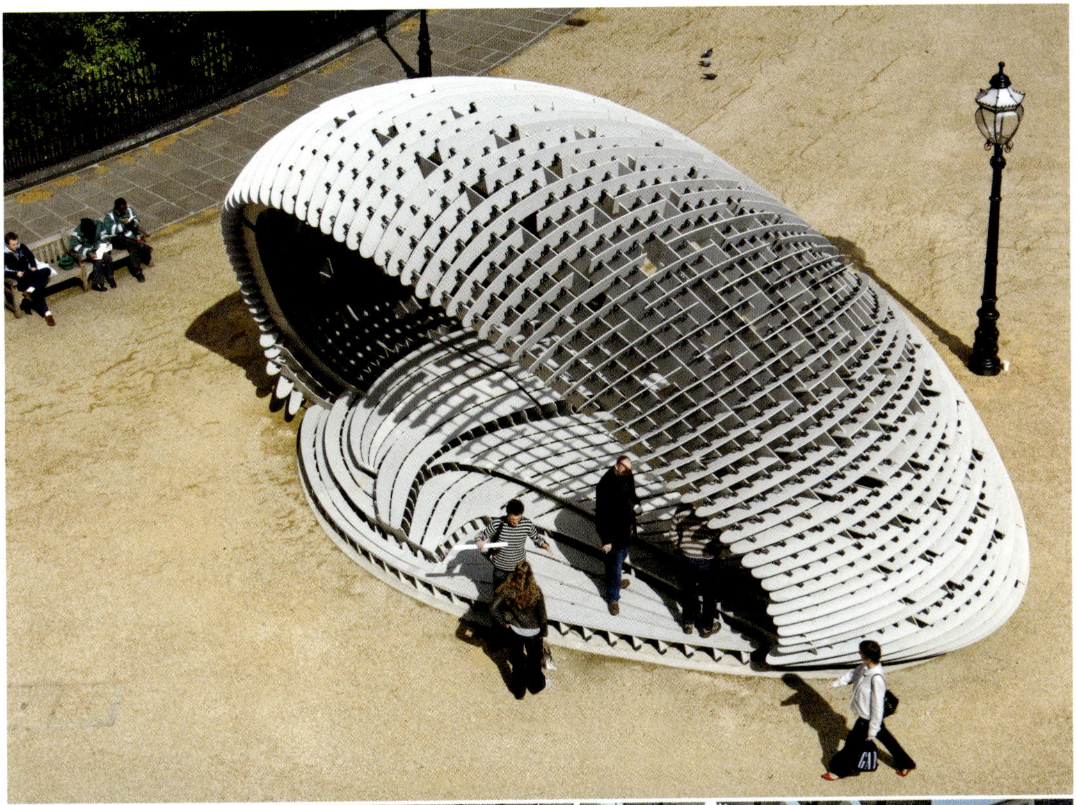

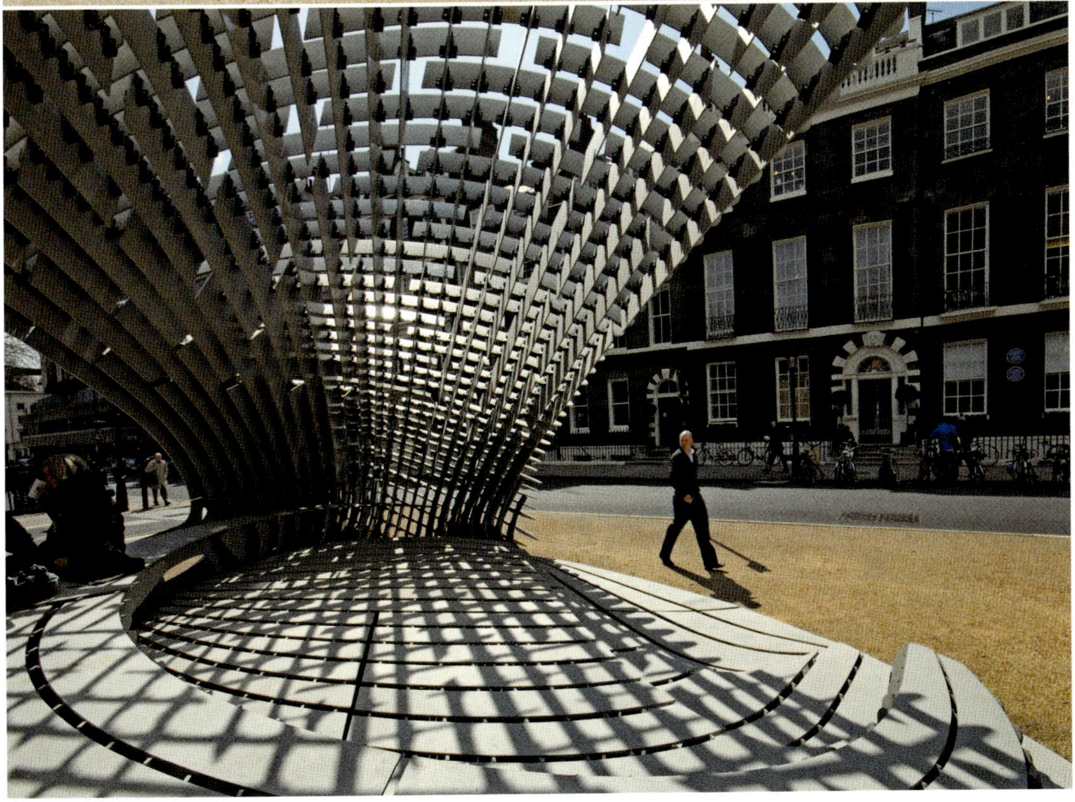

p.art

How can a structural engineer adapt its business systems and organisational procedures in a way that encourages collaboration to be a proactive and diverse engagement, rather than a passive mechanistic adherence to the traditional design process? The approach at AKT has many strands and, as with all complex systems within business and nature, optimum performance is only achieved when each strand works in harmony and with fluidity. One such strand within AKT is p.art, the parametric applied research team.

The aim within p.art is to bring together designers from a variety of backgrounds such as architecture, structural analysis, computer science, forensic analysis, and 3-D graphic designed visualisation and animation. This array of cross disciplinary skills and training is then utilised to find and develop a toolkit of design approaches and methodologies which can be brought to bear on specific project design challenges.

The work within p.art hopes to push the discipline of structural engineering towards a more pluralistic approach by taking advantage of skills from a wider range of design disciplines to enhance the traditional work of the structural engineer.

Plastic opacity
Concrete workshop

p.art

Concrete is as concrete doesn't
MASSUMI

Recent developments in concrete technology, such as high strength concrete and self-compacting mixtures, have improved both its strength and the method by which it is processed. These new properties are bringing about a different level of inspiration to architecture students and practitioners alike by generating new possibilities for concrete usage. These possibilities are much more than technical solutions to design ambitions, whose motivations originate elsewhere. Explorations of concrete's inherent qualities such as mass, weight, density, strength and durability have already lead to innovative applications. Even newer possibilities are created once the very nature of concrete, its opacity, is challenged. The addition of transparency to the list of concrete's properties, which include plasticity, would constitute a huge advancement for concrete technology.

Various developments are shifting our existing notions of transparency and lightness in architecture. Increasingly rigorous physical (or environmental) demands will continue to reduce the surface area of glass used in building, but advanced technologies offer alternative means of preserving transparency. Computing power allows us to identify structural 'cold spots' which can be 'dematerialised', offering seemingly unlimited techniques for generating form. The move from a 'material transparency' towards a 'spatial transparency', in which formal issues such as depth, void and matter meet with material properties like texture, weight and solidity, can now offer experiences and interpretations of transparency that are generated by the opacity of the material. Paradoxically, exploiting concrete's property of opacity offers the potential to experience and increase transparency, but it is a transparency in a relative, rather than an absolute, sense.

The plastic characteristics of concrete can allow for 'free' transformations, while efficiently resolving structural and physical demands. That is the plastic characteristics which can range from fluid through to solid, and can combine with concrete's more traditional characteristics of mass and resilience in order to aid the production of complex forms. *Plastic opacity* infers a spatial transparency, opening up to intricate engagements such as shadow and light, tactility and relief. This leads to the introduction of techniques such as weaving, punching and folding. *Plastic opacity* departs from the realm of the purely visual, and steps into programmatic, environmental and physical aspects, whilst also touching upon specific experiences or spaces, and contexts. There is also further scope for exploring other architectural issues as well.

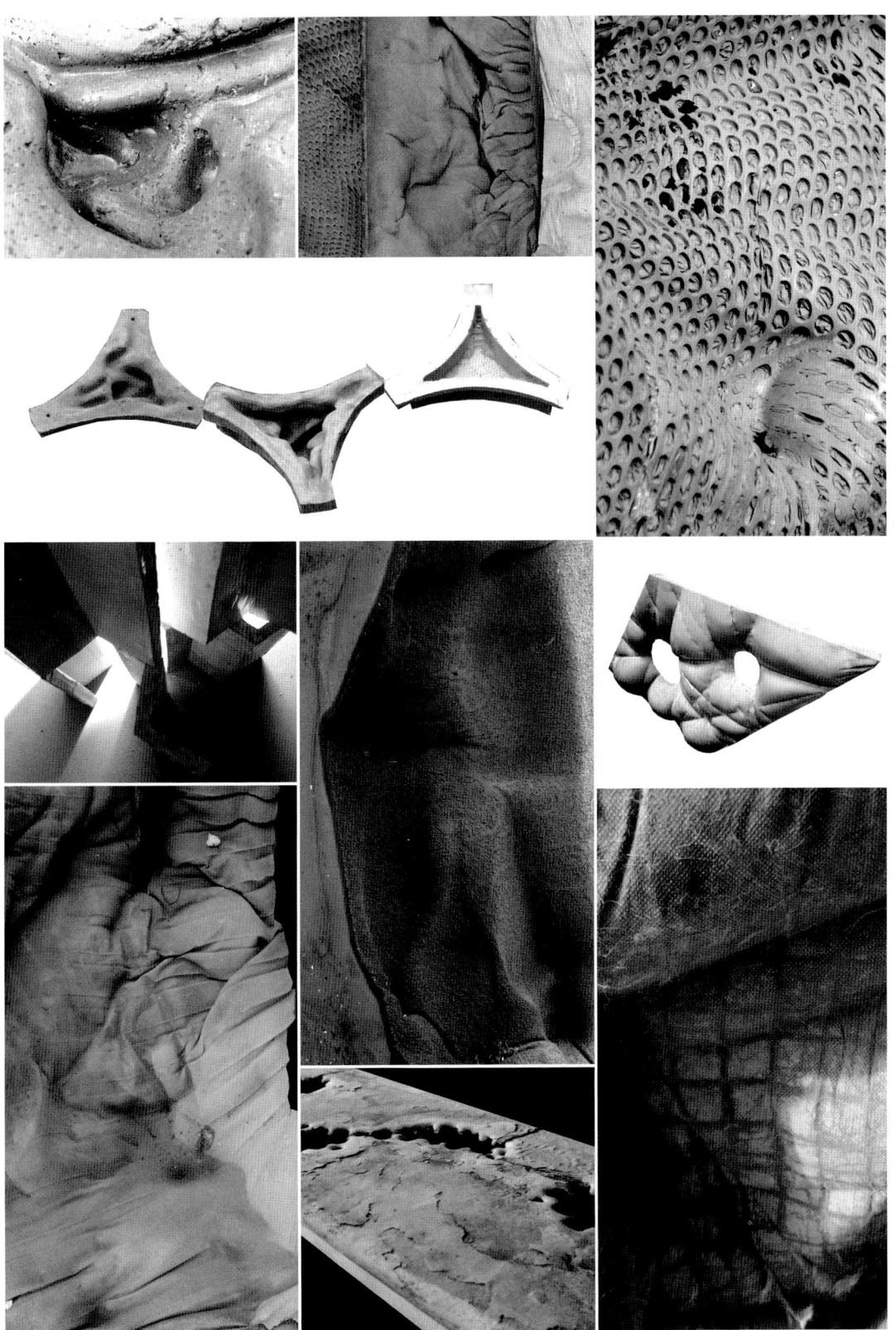

The discipline of design demands a reciprocal relationship that can move from idea to materiality and vice-versa. The dual notions regarding plastic-OPACITY tap directly into some of the basic properties of concrete. Similarly it offers contextual, theoretical and pragmatic design considerations that are seemingly contradictory. This perhaps unnerving or slightly confusing quality needs to be imaginatively resolved by all entrants. Insights and interpretations that may very well differ completely from presented notions on plastic-OPACITY are welcomed and expected.

This competition (is it a workshop or competition? seeks to investigate, through research and design, any notion of plastic-OPACITY in or with concrete. It asks participants to embrace and explore opportunities implied by the dual and combined qualities of plasticity and opacity by allowing the pluralistic and phenomenal implications of both. Results of these explorations have to be presented through proposals that are 'design-led' (be it architectural, structural or otherwise) to reveal their relevance and merits by application. The proposals may range from objects, furniture, buildings and architectural details, to housing, landscape interventions and other large-scale projects. 'Traditional' design criteria such as programme, location, context, scale and so on, may be added freely by participants in order to structure their research and enhance the potential of their application. These can be derived from recent school projects, as the competition aims to blend with current curricula.

The concluding event of the master class was all the more special because of its venue, the Bauhaus in Dessau. It was no coincidence that the event took place in the same year the institution celebrated its eightieth anniversary. For all participants and visitors it was a wonderfully inspiring experience to work, eat and, for some, to stay in a building that is an important piece of world heritage.

For seven days a total of 43 students from eight countries, but of several nationalities, designed and discussed their concrete contributions to the theme. The p.art team with the organisers called for innovative applications and solutions in seven different assignments. Among the presented projects were swimming isles, mannequins as fountains, and garbage columns that addressed environmental issues. After the initial draft designs the formwork was made with timber and foam under the experienced guidance of Guido Lau, head of the wood workshop at FH Anhalt. Students also went to department stores and building markets to buy materials for their various concrete structures. Even the rubbish was taken from garbage bins to put into the formwork. The results published in "plastic-OPACITY" book, the whole event was sponsored by eight European cement and concrete organisations.

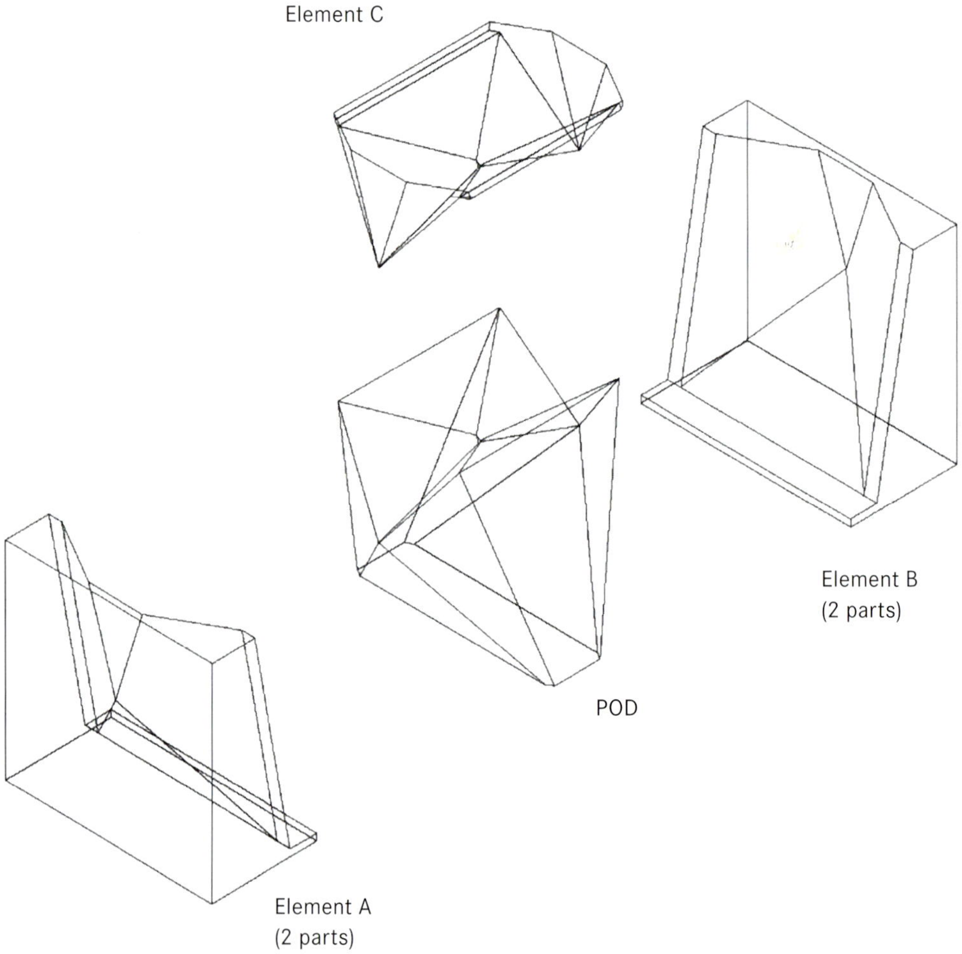

Element C

Element B
(2 parts)

POD

Element A
(2 parts)

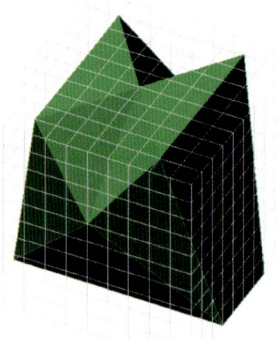

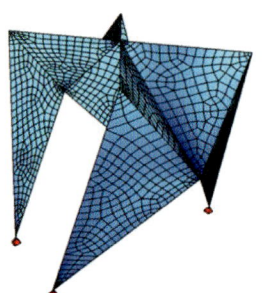

Fibrous concrete
Istanbul workshop

p.art

AKT collaborated with two architectural firms, ecoLogicStudio and Tuspa, in running a pair of workshops in the spring and autumn of 2007 that explored the potential of fibrous concrete. One workshop took place at the Technical University of Istanbul and the other at the Architectural Association, with students from both institutions participating. One goal of this initiative is to stimulate innovative ideas for a pavilion to mark Istanbul's period as European City of Culture in 2010.

Fibrous concrete has been created using thin tubes of concrete that can twist and flow independently, but still interact to ensure stability. Formed by pouring concrete into plastic pipes, there are two basic options for generating adequate structural strength. One is to treat the pipe as permanent formwork, and to introduce fibres into the concrete which would have to be high-strength and self-compacting.

The other is to create a bond between the plastic and concrete so they act as a composite member.

The workshops revealed that interaction between the concrete tubes is the key to making a usable structure from this idea. Provided the tubes come together at points of maximum stress, they can separate into sinews of very narrow diameter; a counter-intuitive effect in concrete. It is clear that such a plastic/concrete composite could work, even if it may not be suitable for the pavilion.

Branching Structure Prototype

Plan

Elevation

bundeling eLS

group: bundling-H

group: branching-V

group: weaving-H-1

group: weaving-H-2

group: weaving-V

Credit of the projects

PECKHAM LIBRARY

Location
Peckham, London

Year
1999

Architect
Alsop Architects

Client
London Borough of Southwark

Contractor
Sunley Turriff Construction Ltd

INSTITUTE OF CELL AND MOLECULAR SCIENCE

Location
London

Year
2006

Architect
Alsop Architects

Client
Queen Mary College, University of London

Contractor
Laing O'Rourke plc

BELGO ZUID

Location
London

Year
1999

Architect
Foreign Office Architects

Client
Belgo

Contractor
Osbourne

MONSOON VILLAGE

Location
Nothing Hill, London

Year
2007

Architect
Allford Hall Monaghan Morris

Client
Hougue Limited

Contractor
Laing O'Rourke

ELIZABETH HOUSE

Location
Waterloo, London

Year
2005

Architect
Foreign Office Architects

Client
P&O Estates

PHÆNO SCIENCE CENTRE

Location
Wolfsburg, Germany

Year
2005

Architect
Zaha Hadid

Client
City of Wolfsburg

Contractor
E Heitkamp GmbH

SOUTHWARK JUBILEE STATION

Location
Southwark, London

Year
1999

Architect
MacCormac Jamieson Prichard / Alex Beleschenko

Client
Jubilee Line

Contractor
Seele (UK) Ltd

HUTCHINSON 3G

Location
London

Year
2002

Architect
Foreign Office Architects

Client
Hutchinson 3G

Contractor
Optima

NAPLES HIGH SPEED TRAIN STATION

Location
Naples, Italy

Year
2003-2008

Architect
Zaha Hadid Architects

Client
TVA

Contractor
TBC

JOHN LEWIS LEICESTER

Location
Leicester, UK

Year
2008

Architect
Foreign Office Architects

Client
John Lewis Partnership / Hammerson

Contractor
Sir Robert McAlpine Ltd

LONDON SCHOOL OF ECONOMICS

Location
London

Year
2001

Architect
Foster & Partners

Client
London School of Economics

Contractor
Bovis Lend Lease Ltd

WALBROOK SQUARE

Location
London

Year
2006-

Architect
Atelier Foster Nouvel

Client
Stanhope plc

Contractor
Bovis Lend Lease Ltd

LITTLEHAMPTON CAFÉ

Location
Littlehampton, UK

Year
2007

Artist
Thomas Heatherwick Studio

Client
Brownfield Catering

Contractor
Littlehampton Welding Ltd

SOUTHWARK GATEWAY

Location
Southwark, London

Year
1999

Architect
Eric Parry Architects

Client
London Borough of Southwark

Contractor
Maunsells

LONDON BRIDGE SCULPTURE

Location
Southwark, London

Year
1999

Architect / Artist
KSS Architects / Lucien Simon

Client
The CIT Group

BBC MUSIC CENTRE

Location
London

Year
Competition 2003

Architect
Foreign Office Architects

Client
BBC

FUTURE HOMES

Location
Travelling exhibition

Year
2000

Architect
Foreign Office Architects

Client
Confidential

MILAN FAIR TOWER

Location
Milan, Italy

Year
2012

Architect
Zaha Hadid Architects

Client
Citylife

Contractor
Bovis Lend Lease Ltd

BUSAN TOWER

Location
Busan, South Korea

Year
2006

Architect
Foreign Office Architects

Client
Solomon Group

QUEEN MARY STUDENT VILLAGE

Location
London

Year
2007

Architect
Feilden Clegg Bradley

Client
Queen Mary College, University of London

Contractor
Laing O'Rourke plc

HEELIS NATIONAL TRUST

Location
Swindon, UK

Year
2006

Architect
Feilden Clegg Bradley

Client
National Trust

Contractor
Moss Construction

BANKSIDE PAVILION

Location
London

Year
2005

Architect
Zaha Hadid Architects

Client
Architectural Foundation / Land Securtires

SOUTHERN RIDGES

Location
Southern Ridges, Singapore

Year
Competition 2004
Construction 2008

Architect
IJP Architects / RSP Architects Planners & Engineers (Pte) Ltd

Client
Urban Redevelopment Agency

Contractor
Evan Lim & C Pte Ltd

STRAND LINK BRIDGE

Location
London

Year
2005

Architect
Future Systems

Client
Land Securities

DRL TEN PAVILION

Location
London

Year
2008

Architect
Alan Dempsey & Alvin Huang

Client
Architectural Association

Contractor
Rieder & Co

AKT employees (2008 Present / Past)

Adam Redgrove
Adam Thomas
Adiam Sertzu
Adrian Power
Albert Williamson-Taylor
Alessandro Margnelli
Alex Hanna
Alex Johnson
Alex Lambrou
Alisina Poya
Alistair Hall
Alvaro Gonzalez
Andrew Pottinger
Andrew Rice
Andrew Ruck
Andy Murray
Ania Wilk
Ann Gray
Anna Blocher
Anna Mierzejewska
Annette Miles
Aoife Bloomer
Avynash Sithanen
Azeem Ramzan
Baktasch Spartak
Belinda Fenney
Ben Butler
Ben Hubble
Ben Lewis
Ben Parry
Ben Whitehead
Bhavesh Varsani
Billy Bilkhu
Birgit Muller
Bruno Guiomar
Carla Ferris
Carlo Diaco
Carmen NG
Carol Chue
Carolina Lameiras
Caroline O'Sullivan
Chanda Powell
Charlie Brandon
Cherry Eagleson
Chloe Faram
Chris Stobbart
Christian Tygör
Christina Hsieh
Christopher Blust

Christos Mitsarakis
Clare Lowe
Clemens Neugart
Colin Apel
Colin Clark
Conrad Coombe
Csaba Kurczina
Cyril Picard
Damien Flahaut
Daniel Spratling
Danise Bartlett
Danny Hambling
Dave Rayment
David Heeley
David Sharples
David Watson
Debbie Wood
Declan Collier
Dee Murphy
Deniz Sutlu
Desrene Dacres
Deyan Marzev
Didier Prongué
Dimitris Linardatos
Dimitris Tousiakis
Dimtrios Karyipidis
Djordje Stojanovic
Dominic Wind
Dorothy Laskowski
Drew Chapman
Duncan James
Ed Moseley
Edoardo Tibuzzi
Efsevia Iliopoulou
Elena Marshall
Eleni Axioti
Emanuel Bringer
Erik Dirdal
Ewa Ciejka
Fabio Bonin
Fabrizio Fortunato
Federica Ariu
Fung-Yin Tsui
Gabriel Sanchiz
Gary Jones
Gary Lynch
George Christou
Gerry O'Brien
Gillian Simpson

Gillian Towns
Gurvinder Plahay
Guy Macleod
Hanif Kara
Harji Patel (Sunny)
Henrich Vitalos
Henry Travers
Hina Solanki
Hirji Patel
Hong Ong
Hugh O'Neill
Ian Bleakley
Ian O'Keeffe
Ian Waddingham
Ieronymos Chochlakas
Ifeanyi Oganwu
Izabela Zawadzka
Jake Ivory
James Barry
James Souter
James Wells
Jamie Whitehead
Jan Friedlein
Jane Brown
Janine Segal
Jason McNee
Jay Harrison
Jay Kerai
Jennifer Burton
Jenny Saunders
Jeremy Gilmore
Jessica Brew
Jim Dunn
John Gerrard
John Risley
Jonathan Onslow
Jordan Brandt
Jose Sarriequi
Joseph Barnett
Joseph Kiafuca
Jugatx Ansotegui
Julian Birbeck
Julie Hynes
Juliette Blisset
Karen Plant
Karin Glöckle
Karin Gockle
Kate Ball
Kate Eyre-Maunsell

Kate Jones
Kate Llewellyn
Katya Boudjelloud
Kelly Aufojul
Ken Wong
Kevin Bennett
Kevin Clark
Kin Wah Cheung
Kristina Kurbalja
Kyron McKay
Lara Giannetto
Laura Saunders
Lee Parker
Lee Wingate
Liesle Brookman
Lina Martinsson
Luana Elford
Mairead Fitzpatrick
Marcin Golebicki
Marco Cerini
Marco Vanucci
Marek Klukowski
Margarida Alves
Maria Georgiades
Maria Giannetto
Marie Joiret
Mario Vounatsos
Marlon Otterwell
Marta Galinanes-Garcia
Martha Enthoven
Martijn Veltkamp
Mary Jones
Masoud Daroghehkaze
Matthew Beth
Matthew Davies
Matthew Phillips
Matthew Phillips
Mei Chan
Melvyn Perera
Michael Barnes
Michael Crabb
Michael Duff
Michael Hartley
Michael Hitchens
Michael Lam
Michael Schumacher
Michelle Chan
Michelle Davies
Montu Vadgama

Naomy Watanabe
Nicola Carniato
Nigel Barker
Nik Devlin
Oliver Bruckerman
Omar Diallo
Pablo Beale
Panagiotis Michalatos
Paul Dwyer
Paul Earwaker
Paul Fuller
Paul Griffiths
Paul Hutter
Paul Scott
Paul Shinkwin
Paul Solomon
Paula Silva
Paulina Oduro
Pavlina Linhartova
Peter Evans
Peter Grinyer
Peter Whyatt
Phillipa Church
Pierre Trufer
Pippa Driver-Davidson
Pritpal Sahota
Rachel Cleary
Raymond Lau
Regina Cheng
Reuben Brambleby
Ricardo Baptista
Ricardo Bittini Miret
Richard Cerfontyne
Robert Clark
Robert Partridge
Robert Spink
Robin Adams
Rohini Char
Rosemary Woods
Ross Revers
Russell Doughty
Russell Henderson
Sabina De Jesus
Sacha Zeilhofer
Samanthan Roache
Samik Narotam
Samuel Ko
Sara Osbaldeston
Sawako Kaijima

Scott Hull
Sebastian Khourian
Sharon So
Simon Chong
Simon Holmes
Simon Jewell
Sofia Effraimiadou
Sophie Treillard
Stefan Albrecht
Stefan Piasecki
Stefan Reuther
Stefano Strazzullo
Stelio Papastylianos
Steve Hood
Steve Nicolaou
Steve Toon
Steven Tallis
Stuart Hammond
Stuart Maddock
Stuart Sagar
Stylianos Papastylianou
Susan Campbell
Susan Mantle
Terry Colley
Thomas Papakonstantinou
Thomas Reynolds
Thomas Whitworth
Thomas Wommelsdorff
Tiana Andre
Tim Pye
Toby Harling
Tom Dowdall
Tommy Wu
Tony Barrett
Urszula Markiewicz
Valentina Galmozzi
Vivianna Romani
Wah Szeto
Warren Sullivan
Yash Kodai
Yuki Oh
Yunna Zhang
Zainab Taylor-Camara
Zara Shariff

Design Engineering AKT Adams Kara Taylor

Edited by
Hanif Kara
Adams Kara Taylor

Published by
Actar

Editorial Concept
Michael Kubo

Editorial supervision
Tomoko Sakamoto
Toby Harling
Albert Ferré

Collaborators
Jessica Brew
Mairead Fitzpatrick
Jan Friedlein
Mimi Zeiger
Irene Hwang

Book design concept
Sandra Neumaier

Design supervision
David Lorente @ Actar Pro

Cover Design
Ramon Prat

Production
Actar Pro

Digital Production
Carmen Galán
Lea Linares

Printing
Ingoprint S.A.

Photographs credits
AKT
except:
Roderick Coyne (Peckham Library)
MVS (Institute of Cell
and Molecular Science)
Valerie Bennett (Belgo
and DRL Ten Pavilion)
Tom Soar (Monsoon Village)
Klemens Ortmeyer
(Phæno Science Centre)
Peter Durant / arcblue.com
(Southwark Jubilee Station)
James Winspear /
View and Paul Rattigan
(London School of Economics)
Trevor Spiro
(London Bridge Sculpture)
Peter Cook / View
(Queen Mary Student Village)
Dennis Gilbert /
View and Heliview
(Heels National Trust)
IJP and MHJT (Southern Ridges)
AA Students (Plastic-Opacity and
Fibrous Concrete workshops)

Adams Kara Taylor part of
WYG Engineering Ltd

AKT would like to thank the many
clients, consultants and collabo-
rators whose projects we have not
been able to include on this occa-
sion. Our special thanks go to the
following contributors to the book:
Simon Allford
Dr. Tim Anstey
Prof. Mohsen Mostafavi
Patrik Schumacher
Prof. Michael Speaks
Alejandro Zaera-Polo

All rights reserved
© of the edition, Actar, 2008
© of the text, AKT and
their authors
© of the photographs,
their authors
© of the graphic documentation,
AKT

ISBN: 978-84-96540-66-8
DL: B-42303-2008

Printed and bound
in the European Union

Distribution
ACTAR D
Roca i Batlle 2-4
E-08023 Barcelona
Tel: +34 93 4174993
Fax: +34 93 4186707
office@actar-d.com
www.actar-d.com

Distribution USA
Actar Distribution, Inc.
158 Lafayette St., 5th Floor
New York, NY 10013 USA
Tel: +1 212 966 2207
Fax: +1 212 966 2214
officeusa@actar-d.com
www.actar-d.com